INTRODUCTION TO STATISTICS

Revised Edition

Introduction to
STATISTICS
Revised Edition

Robert Fried
Hunter College

GARDNER PRESS, INC., New York

Distributed by HALSTED PRESS
Division of John Wiley & Sons, Inc.

New York   Toronto   London   Sydney

This edition published by arrangement with Oxford University Press.

GARDNER PRESS, INC.
32 Washington Square West
New York, New York 10011

Distributed solely by the HALSTED PRESS Division of John Wiley & Sons, Inc., New York

Library of Congress Cataloging in Publication Data
Fried, Robert, 1904-
    Introduction to statistics.

    Bibliography: p.
    Includes index.
    1. Statistics. I. Title.
HA29.F72   1976   519.5   76-20558
ISBN 0-470-15184-6

To Paul, Steven and Dennis.

PREFACE

This text presents concepts and methods in statistical analysis for the typical undergraduate student who has a secondary school level of training in mathematics. My experience, reinforced by my students, is that their difficulty in understanding statistics is verbal and not mathematical: statistics has to be explained carefully. And that is what this text is all about.

The general organization focuses on the elementary probability nature of inferential statistics procedures. It is logically presented to follow a reasonable sequence of content and level of difficulty. This text features a one symbol system consistent with all related concepts, an elementary introduction to new data population, and a chapter on trend analysis. The numerous calculation examples are worked out in detail and the answers are supplied. An alternate set of examples are available in the *Instructor's Manual* available to those who have adopted the text on written request from the publisher.

The revised edition has been refined as a result of its use by a number of individuals since the first edition was published. It is important for the student to discover and understand the concepts which underlie statistics and to practice these concepts as soon as they have been learned; the organization of this text lends itself ideally to this proposition.

Contents

I. *Introduction*   *1*

The Nature of Empirical Observations   *2*
    Discrete and Continuous Data   *3*
    Scaling Variables   *3*
Mathematical and Statistical Proof   *5*
Populations and Samples   *7*
    Populations   *8*
    Samples   *8*
General Rules   *9*
    Quantitative Symbols   *9*
    Operations Symbols   *10*
The Application of Statistical Procedures   *11*
    Descriptive Statistics   *12*
    Inferential Statistics   *12*
    Correlation Analysis   *12*
Problems   *13*
Answers   *14*

II. *Frequency Distributions: I*   *15*

The Range: w   *15*
Frequency Distributions   *17*
    Grouped Data Frequency Distributions   *19*
The Frequency Histogram   *19*

The Frequency Polygon   *21*
    The Cumulative Frequency Polygon   *26*
Proportions and Quantiles   *26*
    Percentiles   *27*
    Deciles   *29*
Problems   *29*
Answers   *30*

III. *Elementary Probability   34*

Determining the Probability of an Event E   *34*
    The Simplest Case   *35*
Conditional Probability   *39*
Permutations   *39*
Combinations   *41*
Discrete Probability Distributions   *42*
The Binomial Distribution   *43*
Problems   *44*
Answers   *45*

IV. *Measures of Central Tendency   47*

The Mode   *48*
The Median   *52*
    Computing the Median from Ungrouped Data   *52*
    Computing the Median from Grouped Observations   *53*
The Mean   *54*
    The Mean Computed from Grouped Data   *58*
Problems   *60*
Answers   *60*

V. *Measures of Variability   63*

The Range (w) as a Measure of Variability   *63*
The Interquartile Range (Q)   *66*
The Variance (S^2) and Standard Deviation (S)   *66*
    S^2 and S Computed from Ungrouped Data   *68*
        Method 1   *68*
        Method 2   *69*
    S^2 and S Computed from Grouped Observations   *69*
Transformation of a Distribution   *70*

Transformation by Adding a Constant   *71*
Transformation by Subtracting a Constant   *72*
Transformation by Multiplying by a Constant   *72*
Transformation by Dividing by a Constant   *73*
Transformation of a Distribution: z *74*
Problems   *76*
Answers   *76*

VI. *Frequency Distributions: II*   *79*

Symmetrical and Skewed Distributions   *79*
The Normal Distribution   *82*
Standard Scores   *86*
The Sampling Distribution of $\overline{X}$ *86*
The Small Sample Statistic t *90*
Problems   *92*
Answers   *92*

VII. *Testing Hypotheses: I*   *96*

The Parametric Assumptions   *98*
The Level of Significance: α *99*
     The Null-hypothesis   *99*
     Type I Errors: α Errors   *101*
     Type II Errors: β Errors   *103*
Confidence Limits   *104*
Point Estimates   *105*
Problems   *106*
Answers   *107*

VIII. *Testing Hypotheses: II*   *108*

Hypotheses About the Population Mean: μ *109*
     Estimating μ when σ Is Known   *110*
          Point Estimate of μ *110*
          Confidence Limits of μ *110*
     Estimating μ when σ Is Not Known   *111*
          Point Estimate of μ *111*
          Confidence Limits of μ *111*
Hypotheses About the Population Variance: σ^2 *112*

Confidence Limits of σ^2 *112*

Hypotheses About the Variance of Two Populations   *113*

Hypotheses About the Mean of Two Populations: μ and μ_2 *115*

Differences Between Means of Independent Populations when σ Is Known and $\sigma_1 = \sigma_2$ *115*

Differences Between Means of Independent Populations when σ^2 Is Not Known   *116*

Differences Between Means of Independent Populations when the Observations Are Paired   *118*

General Recommendations for Rejection of the Null-hypothesis   *120*

Problems   *121*

Answers   *122*

IX. *Analysis of Variance: I*   *124*

One-variable Analysis of Variance   *127*

The Between-groups Sum-of-squares: $\Sigma x_b{}^2$ *130*

The Within-groups Sum-of-squares: $\Sigma x_w{}^2$ *134*

The Total Sum-of-squares: $\Sigma x_t{}^2$ *135*

Summary Table for One-variable Classification Analysis of Variance   *136*

Testing the Difference Between Category Means   *137*

Testing the Significance of Differences Between Pairs of Sample Means   *137*

Testing the Significance of Differences Between Category Means when Variances Are Not Equal   *139*

Testing for Excessive Variability Among Category Means   *139*

Testing the Significance of Differences Between Category Variances   *140*

Problems   *141*

Answers   *141*

X. *Analysis of Variance: II*   *144*

Two-variable Classification Analysis of Variance (Single Observation)   *145*

The Total Sum-of-squares: $\Sigma x_t{}^2$ *147*

The Sum-of-squares for Column Means: $\Sigma x_{bc}{}^2$ *147*

The Sum-of-squares for Row Means: $\Sigma x_{br}{}^2$ *148*

The Residual Sum-of-squares: $\Sigma x_{res}{}^2$ *148*

The Summary Table   *148*

Testing the Effect of Variable I   *149*

Testing the Effect of Variable II   *149*

Two-variable Classification Analysis of Variance (Multiple Observations)   *150*

The Total Sum-of-squares: Σx_t^2 *152*

The Sum-of-squares for Category Means: Σx_b^2 *152*

The Sum-of-squares for Column Means: Σx_{bc}^2 *153*

The Sum-of-squares for Row Means: Σx_{br}^2 *154*

The Residual Sum-of-squares: Σx_{int}^2 *154*

The Within-Groups Sum-of-squares: Σx_w^2 *156*

Analysis of Variance Summary Table   *156*

Testing the Significance of Differences Between Populations   *157*

Testing Significance of Differences Between Variable I Categories   *157*

Testing Significance of Differences Between Variable II Categories   *157*

Testing the Significance of Interaction   *158*

Problems   *158*

Answers   *159*

XI. *Correlation   165*

The Product-moment Correlation Coefficient: r_{XY} *167*

The Correlation Coefficient, r_{XY}, as an Index of Linear Association   *168*

Computation of r_{XY} from Original Observations   *171*

Computation of r_{XY} from a Frequency Table   *174*

Computation of r_{XY} from Standard Scores   *176*

Linear Regression   *177*

The Line of Best Fit: Method of Least Squares   *179*

The Regression of Y on X   *182*

The Standard Error of Estimate   *184*

Testing the Hypothesis that $r_{XY} = \rho = 0$ *186*

Problems   *187*

Answers   *188*

XII. *Correlation: II. Measures of Association   191*

The Biserial Correlation Coefficient: r_b *191*

The Point Biserial Correlation Coefficient: r_{pb} *193*

Computing r_{pb} *193*
The Correlation Ratio: η *194*
The Tetrachoric Correlation Coefficient: r_t *198*
The ϕ Coefficient   *200*
Rank-order Correlation Coefficients   *201*
   Spearman's Rank-difference Correlation Coefficient: ρ_s *202*
   Computing ρ_s *202*
   The Significance of ρ_s *203*
   Kendall's Coefficient of Concordance: w   *204*
Problems   *206*
Answers   *207*

XIII. *Non-parametric Statistics*   *211*

Chi Square: χ^2 *211*
   Computing χ^2 *212*
   Testing Independence in a Two-way Classification Table   *215*
   Correction for Continuity   *217*
   Testing Goodness of Fit   *217*
   χ^2 Test in Two-by-two Tables   *221*
Fisher Exact Probability Test   *222*
The Sign Test   *223*
The Mann-Whitney U Test   *224*
Problems   *227*
Answers   *228*

XIV. *Tests for Trends*   *230*

One-variable Analysis of Variance: Test for Presence of a Trend   *231*
Comparison of Trends in a Two-variable Analysis of Variance   *233*

XV. *Trends in Time-domain Data Distributions*   *237*

Analog Measures   *238*
The Galvanic Skin Response   *240*
   Change in Magnitude and Per Cent Change   *241*
   Conductance Magnitude and Per Cent Conductance Change   *243*
   Logarithmic Transformation   *245*
   Normalized Transformations   *246*
   Measuring Sequential Dependency: λ *249*

The Cardiac Response   *252*
    Estimated Beats per Minute   *253*
    Successive Beats per Minute   *253*
    Per Cent and Relative Per Cent Change   *255*
Normalized Transformation: ALS   *256*
Problems   *258*
Answers   *258*

Appendix   *263*

Bibliography   *297*

Index   *303*

List of Appendix Tables

A.1. Normal Distribution: Ordinate Y at $\pm Z$ and area a between $\pm Z$.

A.2. Cumulative Normal Distribution.

A.3. Percentile Values of the Unit Normal Curve.

A.4. Percentiles of the t Distribution.

A.5a. Distribution of χ^2.

A.5b. Distribution of $\chi^2/_{df}$.

A.6. Critical Values for Cochran's Test for Homogeneity of Variances.

A.7. Per Cent Points in the Distribution of F.

A.8. Critical Values of the Pearson r.

A.9. Critical Values of ρ (rank-order correlation coefficient).

A.10. Values of p/y for Various Values of p (Normal Distribution Unit Area and Unit Standard Deviation).

A.11. Five Per Cent (Lightface) and One Per Cent (Boldface) Points of the Distribution of W.

A.12. Critical Values of U and U' for One-Tailed test at $\alpha = 0.05$ or a Two-Tailed Test at $\alpha = 0.10$.

A.13. Estimates of r_{tet} for Various Values of ad/bc.

A.14. Critical Values of s_i for the Sign Test.

A.15. Squares and Square Roots of Numbers from 1 to 1000.

A.16. Random Numbers.

A.17. Four Place Common Logarithms of Numbers.

I

Introduction

Statistics is the application of mathematical principles to the collection, organization, and interpretation of observations of events in the empirical world around us. Mathematics is a formal logical system consisting of sets of relationships between numerical symbols usually expressed as theorems and postulates. The empirical world of events with which the scientific investigator is concerned consists of sets of observations that suggest relationships expressed as theories and, ultimately, it is hoped, as laws. The investigator employs the mathematics of statistics as a model or analogy of the empirical world. He assumes that a relationship exists between the formal mathematical world of rules, which he has created, and the empirical world of diverse events, over which he seldom has complete control. He assumes that symbol manipulations in the mathematical world have their counterpart in the empirical world, and from this symbol manipulation he makes inferences and predictions about events which have not, as yet, taken place.

Since the investigator is concerned with empirical events, and with observations and predictions that will lead to an understanding of the laws that govern those events, he uses statistics as a tool in making certain kinds of decisions. These decisions are often concerned with the reliability of his observations and inferences. For instance, if a certain event was seen to occur, it might be desirable to determine how frequently it is likely to recur, and under what conditions one may expect it to take place. It is entirely possible that an event may be unique; it may happen only once. Scientific laws cannot be formulated on the basis of unique events.

The decision that an event is likely to occur 75 per cent of the time—let us say, when conditions a, b, and c prevail—may lead to an understanding of the accompanying conditions d and e, which will prevent its occurrence 25

1

per cent of the time. An ideal and seldom-realized aim of science is the exact prediction of the occurrence of events in the empirical world.

The need for statistical procedures in this kind of decision-making results from the fact that the empirical world of observed events is *variable*. Observations of some aspects of the world are seldom, if ever, identical: few persons have the same height or the same weight; carefully machined tool parts may be interchangeable, but they are not exactly alike; a series of dice rolls seldom yields the same consecutive values. Thus, any attempt to understand the organization and nature of the laws that govern, or describe, empirical events must take into account this observed variability. And a thorough understanding of the methods employed in examining the nature of the variability of empirical events is the basis for practically all statistical techniques and tests.

There is no such thing as a *private* science: the investigator's observations must be communicable, so that anyone may duplicate his observations and verify their validity. Because of this essential criterion of science, each major field of interest has a relatively formal way by which its findings are reported to the world. The method by which data are obtained in a study, the way in which they are summarized, and the analysis used to examine them and draw inferences are fairly standard. This standardization has evolved from a need to reduce the ambiguity inherent in verbal communication. It is not entirely eliminated, of course, but it is considerably reduced when formal symbol systems are employed. Statistical procedures aid the scientific investigator by providing a common ground for all investigators to communicate their observations. Standard procedures for the presentation of data in the form of tables, graphs, and charts reduce ambiguity in the communication of often complex sets of observed events.

The Nature of Empirical Observations

Science is based on the planned systematic observation of events. Since mathematics is an invention of man, it is important to understand that the numerical symbols assigned to differentiate between empirical events may not always accurately describe the differences between them. That is, under certain circumstances numerical values assigned to observations may be less than accurate labels for these observations. For instance, while the assignment of numbers serves to differentiate between houses on a street, they do not provide an index of the nature of the differences between the houses. This example is quite elementary to you; you are familiar with using numbers in this fashion. There are, however, instances when the relationship between quantitative differences between events and the numerical labels that identify them is considerably obscured: for instance, is an individual with an I.Q. score of 100 twice as intelligent as one who scores 50 on the same test?

The assignment of numerical values to represent quantitative or qualitative differences between observed events cannot violate the rules of the mathematical model employed: principles of quantity—for example, additivity, subtractivity, multiplicativity, and divisibility—which apply in mathematics, must also apply meaningfully to the events being observed. It is essential, therefore, for the student to understand the relationship between the statistical model and the empirical world, and, in particular, some of its limitations.

Discrete and Continuous Data

Discrete values are often unordered. The numerical values are the labels of mutually exclusive categories. For the sake of simplification, one might assign the value 1 to males and the value 2 to females to identify members of a dichotomous group of organisms. The implication of a quantitative difference in this example is unclear. However, according to the criteria employed in assigning each member of the group to one or the other category, only two categories will be found to exist no matter how carefully one measures.

Continuous values are used to describe events which always yield a greater number of categories as the techniques for observation are refined. A group of persons can be classified in terms of the weight of its members. Some persons may weigh 130, 132, 135 lbs, etc. But upon closer examination, those persons in the category 130 lbs may be found to weigh 130.50, 130.36, 130.075 lbs, etc.

To the scientific investigator, the collection of these data is essential to two goals described in greater detail later in this chapter: description and inference.

Scaling Variables

Whenever observations belong to at least two mutually exclusive categories, the aspect of the world being observed is called a *variable*. In experimental situations, a variable being manipulated, i.e. an *experimental* variable, is referred to as an *independent* variable. The effect of the independent variable is observed in changes in the *dependent* variable. In behavioral studies, the independent variable is often a set of environmental stimuli varying in quantity or in quality, and the dependent variable is often the response measure.

The quantitative value assigned to an observation of a variable is called the *value of the variable*. The assignment of meaningful values depends upon the proper use of scales. All numerical values used to differentiate between events have logical relationships to each other. The composite of these relationships is called a *scale*. There are four basic kinds of scales: *nominal, ordinal, interval*, and *ratio*.

Nominal scales are composed of numerical values that serve to identify discrete categories, e.g. the numbers constitute labels for the categories and imply no quantitative differences that can be meaningfully handled in terms of numerical operations such as addition, subtraction, multiplication, and division. The numbers on houses constitute such a scale.

Ordinal scales have, in addition to nominal property, rank differences. That is, the numerical values of these scales indicate that there are not only differences between categories but also quantitative differences. An ordinal scale ranks the observations with regard to the extent to which they possess more or less of a given property. The ranks do not, however, indicate *how much* more or less of the property each observation has.

There are events with dimensions that are not readily quantified in terms of precise units. It would be absurd to state that a given painting is twice as beautiful as another, or that in one restaurant the food is one-third as tasty as in another. For this reason, ordinal scales are often applied to observed events that differ along a dimension more nearly qualitative than quantitative, and whose qualitative dimension may be difficult to express quantitatively.

Interval scales are equal-unit scales. That is, the distance between adjacent units on an interval scale is the same irrespective of the magnitude of the adjacent scale units. For instance, the distance between 50° Centigrade and 51°C is the same as the distance between 75°C and 76°C, since it reflects a measurable expansion in a column of mercury.

Interval scales are, however, only relative scales. The amount of energy required to produce a certain change in a column of mercury is different at 50°C than at 75°C. Thus the temperature scale reflects relative rather than absolute changes. In the Centigrade temperature scale, for instance, the freezing point of water is used as the reference point for the scale zero: the measures are, then, relative changes around this arbitrary zero. As a consequence, ratios cannot be formed from values of this scale: 50°C is not half as warm as 100°C. And it is possible to obtain values less than zero: $-10°C$, $-25°C$, for instance.

The *ratio scale* has, in addition to nominal, ordinal, and interval properties, an absolute zero. The absolute zero represents a point such that there is no value which falls below that point: zero represents no less than none of the property being examined. In temperatures, zero degrees Kelvin is an absolute zero. The centimeter scale is another good example of a ratio scale: zero cm represents zero linear extent. The distance between 3 and 4 cm is the same as the difference between 251 and 252 cm, 80 cm is twice as much as 40 cm, and so on.

It is the scientific investigator who assigns numerical values to his observations, and it is essentially his responsibility to make sure he assigns these values in such a way that they reflect the type of information appropriate to the mathematical model. Empirical events can be counted, ranked, and measured. The appropriate scale must be carefully chosen to fit the characteristics of the events being observed. Let us suppose that you are ranking aesthetic preference judgments to certain colors. The ranks are determined to be 1, blue, most pleasant; 2, green; 3, red; 4, yellow; and 5, black, least pleasant. The rank 3.5 would be meaningless, since the relationship which exists between the values 3 and 4 is not reflected by the relationship between the colors red and yellow. Failure to use the appropriate scale in measuring a variable can lead to misleading interpretation of the data obtained.

Mathematical and Statistical Proof

Mathematics is not only a formal system, it is a system of logic. A mathematical proof consists of testing one or more theorems and verifying that there are no exceptions to the rules. When no contradictions exist between sets of theorems, a postulate is proven. Statistical proof is different. Statistics is not concerned with absolute proof. Rather, it is concerned with what is likely to happen *in the long run*. Consequently, testing a statistical hypothesis does not consist in verifying that there are no exceptions to the rule, but rather in determining how many exceptions one should expect in the long run.

The scientific investigator establishes, usually beforehand, a criterion of reliability for his observations and the derived inferences. This criterion is a statement of the number of times he thinks he is likely to be wrong, in the long run, if he accepts his findings as valid. He collects his data and analyzes them to determine whether an event in question occurred, and if so how often it occurred. It is not essential that an event be observed under all manner of conditions. It must, however, have occurred a significant number of times. If an event is seen to occur a significant proportion of the time, the investigator assumes that the event has a probability greater than that which might be expected if the event occurred only through coincidence or chance. This statement provides the basis for the use of statistics in testing scientific hypotheses and in making scientific decisions.

If a mathematical statement is "true," there are no exceptions to the rules. If a statistical statement is "true," the exceptions are relatively rare, perhaps of the order of magnitude of 5 per cent or 1 per cent. A thorough understanding of the methods of statistics permits the investigator to reduce his reliance on chance events and thus improve the accuracy of his predictions.

Let us consider an example which we will discuss in greater detail later. Suppose you toss a coin many times, and record whether a head (H) or a tail (T) occurs each time the coin falls to the table. The mathematical model which describes the probability of H or T indicates that you would expect, in the long run, 50 per cent H and 50 per cent T.

You cannot predict what will happen at each coin toss; since the probability of H is the same as the probability of T, each is equally likely at each toss. You would predict that nothing favors H over T, and therefore the probability should be exactly the same for both events, in the long run. Empirically, you would find that the number of times each event occurs is not exactly 50 per cent H and 50 per cent T. It is probably very nearly 50 per cent for each, but not exactly so. You would probably ascribe the disparity to a chance factor related to random variation in the way the coin was tossed, its distance from the table, etc. A small departure in the empirical observation of events from that expected from the mathematical model would not be considered significant; it would probably be ascribed to unknown chance factors. How much of a departure would be considered significant?

If you tossed coins with a friend and he invariably tossed 75 per cent H, you would soon tend to suspect that he was introducing a variation which made it an unfair game. Your friend is exceeding expected gain. This simple example illustrates the basis for testing statistical hypotheses concerning the nature of certain events; it also illustrates the nature of decision-making processes in scientific investigation. In the coin-tossing problem there are two possible events: H and T. Only one can occur. Either H or T can be observed when the coin is tossed. Since one out of two possible events may occur, we say that the probability, P, is $\frac{1}{2}$, or .50. This statement may be summarized as $P(E) = \dfrac{E}{E's}$: that is, the probability of the *event predicted* equals the event that occurs divided by the number of all *possible events*—including the event that *does* occur.

For any given toss of a coin, we can indicate the relationship between the alternatives and our choice in terms of a decision-making schema:

Alternatives

		H	T
Choices	H	+	−
	T	−	+

There are two ways in which you can guess correctly (HH and TT) and two ways in which you can guess incorrectly (HT and TH), depending on what alternatives actually occur.

All statistical proof is based on this basic probability form: $P(E) = \dfrac{E}{E's}$.

The example above is a simple one, of course, and some of the statistical proofs we will examine in later sections will be considerably more complex and will not appear to be based on $P(E) = \dfrac{E}{E's}$. However, you will see that, barring differences in the nature of the E and the E's, the format remains the same. The complexity of the probability estimation is based on the number of alternative observation categories and the relative frequency with which each is observed in the long run. For example, the probability of selecting any one card from a standard deck is $P(E) = \dfrac{E}{E's}$, or $\dfrac{1}{52}$. In a standard deck there are 52 categories, and one observation in each category. The complexity of the probability estimate increases with the extent to which there is a different number of observations in each of the categories of the variable being observed. For instance, the probability that a given person has a height of 5′ 9″ is not readily determined, because each height category, i.e. 5′ 1″, 5′ 2″, etc., contains a different number of observations.

The scientific investigator is concerned with determining the probability that certain events occur more frequently than might be expected on the basis of chance. What he often does is perform his observations under *controlled conditions*, thereby changing and reducing the effect of a chance occurrence. Determining that an event occurs with a probability greater than that expected on the basis of chance fluctuations is a *statistical proof*.

Populations and Samples

The scientific investigator is concerned, generally, with observation of the properties of the empirical world consisting of a *universe* of variable observations. A *statistical universe* consists of all possible observations on a specific group. This group is defined by a specific characteristic. A statistical universe might consist of the observations of height of all college students, or those of the students in your class. If there are observations hypothetically possible which cannot be made available to the investigator, the universe is said to be *infinite*. If all observations are possible within a reasonable time period, the universe is said to be *finite*. It is not the size of the universe that determines whether it is finite or infinite, but the availability of the observations that can be made on the universe.

Populations

In the behavioral sciences, the universe of observations of interest to the investigator is generally referred to as a *population*. The characteristic that defines the population is called a *parameter*. In the example above we introduced two populations: all college students, and the students in your class. The parameter with which we were concerned was height.

Any summary measure of a population is called a *parametric value*. For instance, the average height of the students in your class, the class *mean* height, is a parametric value.

Since scientists are concerned with the general applicability, or generalizability, of their research findings, they are concerned with statements they can make about populations of events. Making valid inferences about populations is the highest order of generalizability possible. In some cases, however, a population cannot be examined entirely: it may be infinite, or it may be such a large finite population that it is impractical to handle all possible observations. The investigator must then use a shortcut method: this consists of approximating the population by studying a *sample* of observations from that population.

Samples

Sampling consists of obtaining observations on a restricted number of all the events that comprise the population. A *sample* is a finite portion of the population. There are a number of different ways in which samples may be obtained.

Random sampling is a process which assures that no single observation is selected for inclusion in the sample in such a way that it is favored over any other observation. Each observation in the population has an equal chance of being included in the sample. When one or another observation is favored for inclusion in a sample, the sample is not representative of the population; it is then said to be *biased*.

A random sample of adequate size will differ from the population by very little. In many populations there are values of the variable that occur more frequently than others: among college students, the height 5′ 8″ will occur more frequently than the height 4′ 6″. In a random sample, we would not only expect the observation 5′ 8″ to occur more frequently than the observation 4′ 6″, but we would also expect these values to occur with the same frequency as they do in the population. Any difference between the sample and the population, when random sampling is employed, is said to be due to *sampling error*.

Under certain conditions, sampling is bound to be biased. This is particularly true if you know something about the constitution of a population with a small number of categories, in which values may be observed. Suppose you are concerned with consumer preference for a given shaving lotion. Ordinarily, let us say, 75 per cent men and 25 per cent women purchase this product. It would hardly make sense to sample at random in a department store, since, first, there are likely to be more women than men there at any given time, and, second, even if that were not the case, at best you could expect 50 per cent men and 50 per cent women. This hardly resembles the consumer population, and inferences might be irrelevant. However, a sample can be constituted in such a way that it includes 75 per cent men and 25 per cent women. Such a sample is said to be a *stratified sample*.

We could make the assumption that all continuous variables are, for practical purposes, discrete, since we can seldom take all possible categories in a continuous variable and observe them. Following this reasoning, random samples are then really stratified samples with large numbers of categories. However, this is not common usage: stratification usually applies to discrete variables, particularly when the number of categories is relatively small.

General Rules

It is essential that the student familiarize himself early in the course with some of the vocabulary common in statistics. Very often the failure to comprehend statistics is not associated, as many students believe, with a lack of aptitude for mathematics (specifically arithmetic), but rather with a failure to comprehend the required operations because of a language barrier. Without a thorough understanding of the symbols used the simplest formula is incomprehensible.

It would be well if the student made an effort to commit to memory all those symbols with which he is not now familiar. These symbols are shorthand expressions for communicating information, and they generally convey either quantitative data or statistical operations.

Quantitative Symbols

Hypothetical quantities may be represented by letters of the alphabet: a, b, c, . . ., etc. Typically, an observation is represented by the letter X. When there is more than one observation in a set, and one specific observation is indicated, a subscript is used: X_1, X_2, X_3 . . . X_n indicate the first, second, third observation, and finally, the nth observation, there being n observations in a set.

Under certain conditions, it is irrelevant whether a quantity is positive or negative; we are concerned only with the *absolute* value of that quantity or magnitude. Absolute value of the observation X is indicated by $|X|$.

The upper-case letter X, as we said, represents any observation in the *x variable*. It might represent a given height, or weight, or the speed of a rat in a maze. When two variables are being examined, the observation in the second variable is represented by the letter Y.

Here are additional symbols and the relationships they indicate:

$a = b$ a *equals* b in magnitude

$a \simeq b$ a *is approximately equal to* b in magnitude

$a \neq b$ a *is not equal to* b in magnitude

$a < b$ a *is less than* b in magnitude

$a > b$ a *is greater than* b in magnitude

$a \leq b$ a *is less than or equal to* b in magnitude

$a \geq b$ a *is greater than or equal to* b in magnitude

The number of observations in a sample is represented by n, but in a *population* it is N. When a sample consists of more than one category, the number of observations in each category is the category *frequency* and is represented by f; the total number of f's is equal to n. Suppose you have a sample of persons consisting of 35 males and 65 females. There are two categories, male and female, being considered. The frequency in the first category, males, is 35, and in the second category, females, is 65. The sum of the frequencies is 100, or n.

Operations Symbols

There are symbols which do not represent quantities, but operations. These symbols are, essentially, instructions. You are, of course, quite familiar with $+$, $-$, $\times$, and $\div$. These operations may be represented in several different ways. Note the following additional symbols for these common operations: the symbol Σ (Greek sigma) represents the *sum* of a set of observations (X) and indicates the operation of addition:

$$\Sigma X = X_1 + X_2 + X_3 + \ldots X_n$$

The statement above is read: "the sum of X equals X_1 plus X_2 plus X_3 plus $\ldots X_n$", and indicates that the sum of the X values results from the addition of all X's, there being n of them.

The multiplication of the quantities a and b can be indicated as ab, and division of these quantities can be indicated as a/b, or $\frac{a}{b}$.

You will note that Greek letters represent parametric values, while English letters represent statistics: for instance, the mean or average of a population is represented by μ (Greek mu), while the mean of a *sample* is represented by $\bar{X}$ (read X-bar).

Additional symbols will be introduced in later sections. They will represent, generally, either specific data or operations in context with the topic examined.

The Application of Statistical Procedures

The application of statistics is not restricted to any particular field of investigation. Statistical procedures are equally important to the census taker, who is concerned with the general description of populations he studies, the scientist, who is concerned with the attributes of special populations he is trying to understand, and the layman, who is continuously exposed to statistical terminology in mass communication media.

The census taker may deal with *populations* and *subpopulations*. His task is to determine the relevant parametric measures which describe the population, and the proportion of the subgroups that are contained in each subgroup category. For instance, he might wish to determine what proportion of the population consists of males and what of females. He might wish to know what proportion of males are married, the proportion of the subpopulation of married persons who have children, and how many children they have. In addition, he might wish to determine, for the purpose of tax revenue estimation, what the yearly income of each member of the population is. Since there are a large number of categories which can be used to represent actual, exact income, he might *group* persons into smaller salary range groups. These categories might include salary range from $0.00 to $2500, from $2501 to $5000, from $5001 to $7500, etc. These categories vastly reduce the amount of information which the census taker must handle when analyzing his data. In the process of reducing the amount of information by grouping categories there is a loss in the total amount of information available, i.e. the individual observations lose their identity when they are included in larger categories. But this loss is considered minor; grouping results in a greater ease in data processing, which compensates for the loss.

The scientific investigator is often concerned with various unique populations. The physicist may be concerned with a population of gas molecules, the biologist with a population of micro-organisms, the horticulturist with a population of apple trees, and the psychologist with a population of learners. These investigators, and many others; have in common the fact that they will make observations on samples drawn from populations with which they are concerned, and will then make inferences about the populations from the sample data.

In daily life we are continuously subjected to information involving "statistics" in one form or another. In television commercials, for instance, we are sometimes told that a given product ". . . is 50 per cent more effective. . . ." We are not told, however, whether this means that the product is 50 per cent more effective than it was previously, or whether it is 50 per cent more effective than a competing product. We are left to make the inference that the increase in effectiveness now makes the product more desirable. It may well be that the effectiveness of the product was only 10 per cent to begin with (it did what it was supposed to do only once in ten times) and the 50 per cent improvement has made it effective 15 per cent of the time. Knowledge of this factor would hardly tend to induce the consumer to buy the product. This example also illustrates what is meant by "... you can make statistics say anything ..." or "... statistics can be made to lie." In reality, statistics *cannot* be made to say just anything—*to those who understand its language*. Understanding statistical processes and methods should help the novice ask the right questions—the questions that will result in his making meaningful inferences based on available information.

There are three broad categories of statistical procedures: *descriptive*, *inferential*, and *correlation*.

Descriptive Statistics

Descriptive statistics is concerned with the enumeration and summary of the observations obtained from samples and populations. The observations may be counted or ranked or presented in graphic form for visual inspection. The data may also be summarized in such a way that a representative value of some kind is obtained. One might wish to know something about the most typical, or representative, value in a sample or population. One might also wish to know the extent to which observations differ from each other in a sample or population, that is, the variability of the observations. The census taker, for example, deals in descriptive statistics.

Inferential Statistics

Inferential statistics are concerned with making meaningful inferences about populations, or from samples to populations. The criterion for an acceptable inference is an index of the number of times, in the long run, you can expect to be wrong if you accept the inference. For instance, you might, at random, select a sample of machine parts coming from an assembly line. You check the tolerances specified and measure the parts to determine the extent to which they fit the specifications or deviate from them. You can now make some inference about the rest, the total population, of the parts made by the operators on that line on the basis of the data in the sample.

Correlation Analysis

Correlation techniques aid the investigator in predicting the values of one variable, say from population X, from the values of another variable, say from population Y. Correlation proper is a measure of the association of the values of the X and Y populations. That is not to say that it is a measure of causality, as is often mistakenly thought. Correlation simply indicates the extent to which one is likely to find a when one finds b. That is not the same thing as causality. Statistically, there is a high degree of association between headaches and the consumption of aspirin. Can one infer that aspirin causes headaches?

Closely related to correlation is regression analysis. This procedure is used to make predictions based on *correlated* data. For instance, given two variables whose values are know to be correlated, such as height and weight, one may predict the average weight for a given height, and the average height for a given weight.

Problems

1. Identify the following variables as being continuous or discrete:
   a. The weight of the students in your class.
   b. The numerical indices used to indicate women's dress sizes.
   c. Linear extents measured in inches.
   d. Speed in miles per hour.

2. Indicate whether the following observations can be measured on a nominal, ordinal, interval, or ratio scale:
   a. Temperature in degrees Kelvin.
   b. Temperature in degrees Centigrade.
   c. Distance in meters.
   d. Military ranks, e.g. Pvt., Cpl., Sgt., etc.
   e. The Michelin culinary index, e.g. ****, ***, **, *.
   f. The I.Q. as measured by the Stanford-Binet test.
   g. The grade assigned to eggs in your supermarket, e.g. AAA, AA, etc.

3. Indicate whether the following are most appropriately subject to mathematical or statistical proof:
   a. $a^2 + b^2 = c^2$.
   b. There is a good chance of obtaining two heads if you toss two coins.
   c. $c = 2\pi r$
   d. $a^2 + b^2 = (a + b)^2$
   e. Fertilizer a is more effective than fertilizer b in making certain plants grow.

4. Which of the following are finite and which are infinite populations?
   a. The school children in district three.
   b. The children in a given class.
   c. Molecules of gas in a vessel.
   d. Running speed of Sprague-Dawley rats in a 15 ft straight maze.
   e. Achievement score of children in a school district on achievement test A.

5. In which situation would you recommend random sampling and in which would stratified sampling be more appropriate?

   a. Sex of consumers choosing between two brands of soap.
   b. Attitude of a given population to our foreign policy.
   c. Average height of children in a school district.
   d. Average salary of homeowners in a residential area.

Answers

1. a. continuous
   b. discrete
   c. continuous
   e. continuous

2. a. ratio
   b. interval
   c. ratio
   d. ordinal
   e. ordinal
   f. interval
   g. ordinal

3. a. mathematical
   b. statistical
   c. mathematical
   d. mathematical (and false)
   e. statistical

4. a. infinite (if you do not specify only those currently present)
   b. finite
   c. finite (though difficult to count)
   d. infinite
   e. finite

5. a. stratified
   b. random
   c. random
   d. stratified

II

Frequency Distributions: I

The scientific investigator is generally concerned with the nature of events in the empirical world. Such events are usually observed values of some variable. Observing a representative sample of the values of a given variable may result in a large number of such values. There may be a dozen observations in a sample, or there may be hundreds, or thousands. In this section, we will be concerned with a sample of 100 observations. The student may assume that they might result from the observation of any variable, let us say variable x. These observations are given in Table 2.1.

Examining the data in Table 2.1 reveals relatively little at first, other than that they consist of numerical values of one and two digits. Further inspection reveals that the numerical values in the data lie between 11 (the lowest value observed) and 35 (the highest value observed). The data may be said to *range* between 11 and 35.

The Range: w

The *range* (w) is one way in which the numerical observations may be summarized. It is an index of the extent of the *spread* of the observations between the empirically determined limits 11 and 35. The range is obtained by subtracting the value of the lowest observation from that of the highest:

$$w = X_h - X_l$$
$$w = 35 - 11$$
$$w = 24$$

The range is a useful index of the spread of the observations in the data since it may be used to distinguish between different sets of data: samples and

15

populations may differ with respect to their range. In addition, variation may be expected in the range of samples independently obtained from a given population. This variation in the range of samples is implicit in the concept of random sampling: if, when sampling, one expects that there will be variation in the observed values from sample to sample, there is no reason to suppose that the extreme values (or limits) of each sample should not vary as well as any other observation within the range in the samples.

24	28	27	31	19
16	20	20	18	22
19	27	16	34	28
29	32	24	17	26
25	30	23	23	13
23	24	26	22	32
26	31	20	22	14
25	17	18	28	20
24	26	26	29	25
31	25	23	33	22
13	24	18	19	25
12	21	20	23	22
24	24	23	35	18
24	21	21	19	23
30	11	17	25	22
21	29	19	20	15
22	16	23	21	24
24	28	14	25	15
21	27	23	21	27
30	22	26	27	22

TABLE 2.1

Looking at the observations again, you will also note that some occur more frequently than others. The values 11, 12, 33, 34, and 35 each occur only once, while the value 24 occurs ten times. The frequency with which each observation occurs is differentially distributed among all possible values of the variable observed.

Suppose you were to observe a series of weights. Before beginning the process, you would have no limit for the values representing observed weights. A weight may be extremely small—nearly zero—or so great that it can only be estimated since no measuring device exists that will accommodate it. Thus, all hypothetical values of weight (weight categories, or weight-scale categories) would range between zero and infinity for all practical purposes. All hypothetical weight-scale values are referred to as "categories" since they may or may not actually be observed.

The range of all categories that might include observed values is not restricted by the observed values, but not all values need be empirically observed. It may well turn out, in observing the series of weights, that some have the same value. And it may turn out that some categories are not observed: there is nothing about a weight scale that prevents persons from weighing five pounds, but the category, five pounds, is not expected to be represented too frequently. Thus a process of observation is always, in a sense, comparative. That is, when one observes something, e.g. weight, distance, etc., one compares observed events with scale categories.

There are some instances when it is not possible to observe certain events with a frequency greater than one per category: in a standard deck of playing cards, the categories constitute a nominal scale. Specific suit and number combinations occur only once in each deck, and all observations occur only once per category. It would hardly make sense to have a nominal scale where the frequency per category is greater than one: this would be like having three children and naming all three John. It should be noted at this point that whenever a series of observations falls on a nominal scale, and the frequency of observations per category is one, the probability that any one observation will be selected from the series by random sampling is *one* divided by the *number of filled categories*, or $\frac{E}{E's}$. As we will see later, the relative complexity of a probability estimate depends upon the nature of the distribution of observations in the categories of the variable scaled.

The 100 observations with which we are concerned do not fall on a nominal scale. And there are some observations that occur more frequently than others. This property of data, that certain hypothetical categories are represented more frequently than others, is called the *distribution* of the data: the observations are distributed among the categories.

Frequency Distributions

One of the important differences between samples and populations of observations is the differential distribution of empirically observed values among the possible categories. Looking again at the data, the nature of this distribution can be determined by determining the frequency with which each observation occurs.

One convenient method for doing this is to set up a *tally sheet*. The tally sheet is a summary sheet consisting of all possible categories within the range of the observed values. There is no point in listing all possible categories between zero and infinity; however, all categories within the range are listed since we also wish to know which have a frequency of zero. Table 2.2 shows a tally sheet of the data in Table 2.1.

Category	Tally	Frequency
11	|	1
12	|	1
13	||	2
14	||	2
15	||	2
16	|||	3
17	|||	3
18	||||	4
19	₩	5
20	₩ |	6
21	₩ ||	7
22	₩ ||||	9
23	₩ ||||	9
24	₩ ₩	10
25	₩ ||	7
26	₩ |	6
27	₩	5
28	||||	4
29	|||	3
30	|||	3
31	|||	3
32	||	2
33	|	1
34	|	1
35	|	1

$$\Sigma f = 100$$

TABLE 2.2

In the tally sheet the data are enumerated and the occurrence of each observation is noted by a vertical line (|) in the appropriate category. The vertical lines are, for convenience, combined into groups of four, with a line drawn across to represent a fifth observation (₩). This makes it possible to determine rapidly, by counting the groups of observations, what the approximate frequency is in each category. Note that the sum of the frequencies is equal to the total number of observations.

Grouped Data Frequency Distributions

When there is a large number of observations, it is desirable to reduce the number of categories by *grouping*. Grouping is accomplished by dividing the range into *intervals*. Typically, the range is divided by a numerical value so that there will be from 10 to 20 intervals.

The tally sheet for the frequency distribution of the grouped data is similar to that for the ungrouped (raw) data. The difference is that the size of the interval in the ungrouped frequency distribution is 1. For the data in Table 2.1, an interval size of 2 provides 13 intervals, as shown in Table 2.3.

Interval	Tally	Frequency
11 - 12	||	2
13 - 14	||||	4
15 - 16	‖‖	5
17 - 18	‖‖ ||	7
19 - 20	‖‖ ‖‖ |	11
21 - 22	‖‖ ‖‖ ‖‖ |	16
23 - 24	‖‖ ‖‖ ‖‖ ||||	19
25 - 26	‖‖ ‖‖ |||	13
27 - 28	‖‖ ||||	9
29 - 30	‖‖ |	6
31 - 32	‖‖	5
33 - 34	||	2
35 - 36	|	1

TABLE 2.3

The Frequency Histogram

The data in Table 2.1 have been employed in the construction of Figure 2.1. This figure is called a *frequency histogram*—a *histogram* for short. The histogram is also sometimes called a *bar-graph* or *bar-chart*. It is a graphic presentation of a frequency distribution; it shows the frequency with which each category is represented in empirical observations.

The scale categories are indicated on the base of the graph (the X axis, or abscissa). With the exception of correlated data, the results of an experimental investigation are presented graphically, so that the categories represent units of the dependent variable. The Y axis, (the ordinate) perpendicular to the base, represents the frequency with which each category is observed. Irrespective of whether the data conform to a nominal, ordinal, interval, or ratio scale, category scale units on the abscissa are, by convention, generally separated by equal distances. The numerical value of each unit on this base is indicated as the *midpoint* of each category—thus the actual limits of the category 14, for instance, are not 14 and 15, but 13.5 and 14.5. This means that the value 14 represents all values ranging from 13.5 to 14.5.

The data in Table 2.1 are shown plotted in histogram form in Figure 2.1.

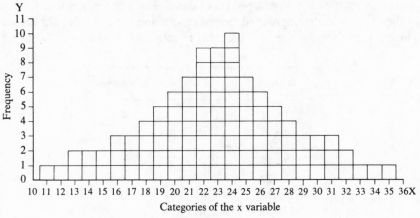

Categories of the x variable

FIGURE 2.1

The Y axis describes the frequency with which the observations occur in each of the X axis categories. The frequency units, also equidistant from each other, provide a range of possible frequencies at least as great as the frequency with which the most frequently observed category is represented.

In the histogram each empirical observation is represented by a *finite amount of space*. The space allocated to the observations is the same for each one of them. Since there is one square space to represent each observation, and the dimensions of this space do not vary with category, the sum of these spaces constitutes an area analogous to the sum of the frequencies, and consequently to the sample size, or population size.

On the tally sheet the individual observations are represented by a |; in the histogram they are represented by a finite bounded space—a square or other area. And, just as it is possible to identify any given proportion of the observations in the tally sheet, so also it is possible to determine any given proportion of the area of the histogram which has as boundaries the base of the histogram and the perimeter of the combined squares representing each observation.

Examine the histogram in Figure 2.2, which represents the grouped data. Again, the sum of the areas representing the individual observations constitutes a figure whose entire area represents the size of the sample or population observed. Irrespective of the number of observations in the series, the area of the histogram represents 100 per cent of the observations.

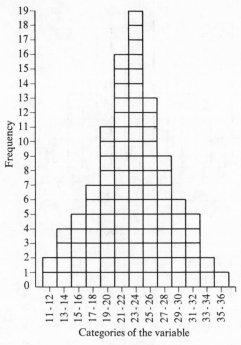

FIGURE 2.2

The Frequency Polygon

Figure 2.3 shows the conversion of a histogram to a *frequency polygon*. The midpoint of each category on the base (X axis) represents the category, and the polygon line connects the midpoints of successive categories at the top of the column. Since the height of the column indicates the number of observations in the category, the intersection of the polygon line and the category midpoint indicates the frequency with which each category is represented in the graph. It is essential to note that the frequency polygon takes into account the implied limits of the data. For instance, the lower limit of category 11 is 10.5 and the upper limit of category 35 is 35.5. And although there is a difference in the shape of the histogram and the polygon, the area in the histogram and polygon are the same.

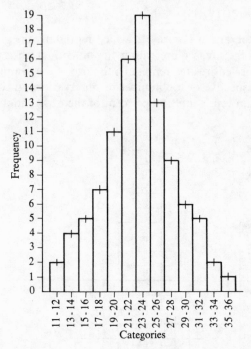

FIGURE 2.3(a)

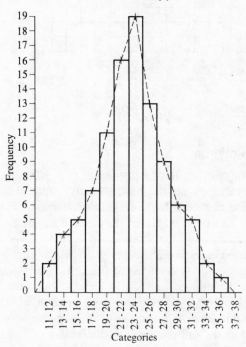

FIGURE 2.3(b)

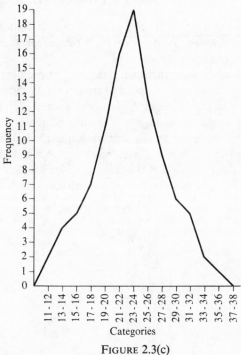

FIGURE 2.3(c)

Beginning students in statistics often fail to understand why it is possible to determine the proportion of the "area under the curve," because they do not realize that the proportion in question is not a proportion of a *polygon* line, but a portion of a closed figure with a finite area that represents all observations in the sample or population graphed. The fact that the area of the histogram and polygon are equal is illustrated in Figure 2.4. This figure is essentially a histogram on which a polygon has been superimposed. The data are those in Table 2.2, *grouped* data.

To simplify the drawing, the categories are not sudivided to show each observation. However, the number of observations in each category is determined by the height of each bar. The bars are measured in frequency units, indicated on the Y axis of the graph. Since the units of the Y axis indicate frequency, or number of observations, the height of the column, or ordinate, also indicates the frequency in that category.

The polygon line superimposed on the histogram appears to cut off certain portions of the histogram. But notice that the line of the polygon also includes in its area portions of the total area which were not previously included in the histogram. These area portions are triangles, and they are indicated as shaded triangles in Figure 2.4. For every portion of a histogram column that was excluded from the polygon structure, a corresponding area was added. The triangles a, a', b, b', c, c', ... n, n', can be demonstrated to be pairs of congruent triangles. And, if you will remember your high school geometry, congruent triangles are equal in area.

There are several other ways in which data may be presented graphically. Every graphic method serves the purpose of transforming data from numerical to spatial representation. The graph is an *analog*, or model, of a numerical frequency distribution, and it maintains a certain degree of integrity of the data from the form of the graph to the distribution of the individual observations. The units on the abscissa (base) represent the categories, and the ordinates at each column represent the frequency with which each category is empirically observed. Co-ordinates of the abscissa and ordinate may,

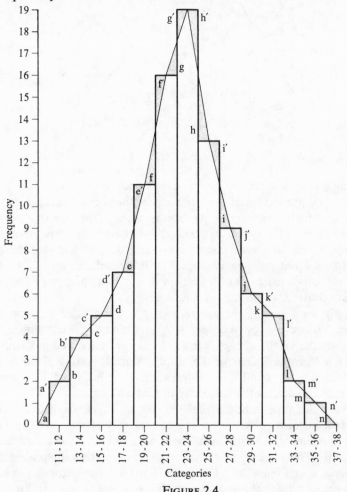

FIGURE 2.4

therefore, be used to divide the polygon so that exact proportions of the area may be obtained. Figure 2.5 shows how the process of obtaining proportions of the area, or of the observations, may be conceptualized.

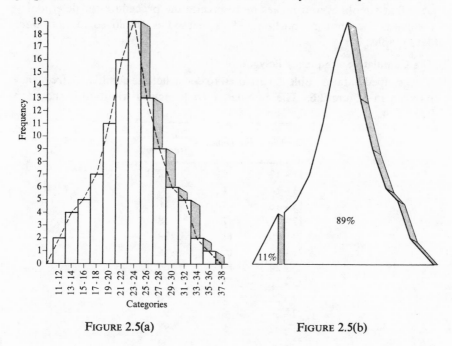

FIGURE 2.5(a)   FIGURE 2.5(b)

Figure 2.5 is a histogram drawn in three dimensions with a polygon superimposed. The polygon is then represented in three dimensions, emphasizing the fact that a graphic representation of the observations is represented by finite area. Let us suppose, for example, that we wish to determine below what point (category) on the base we can expect to find 11 per cent of the observations. The sum of the frequencies in the categories 11 – 12, 13 – 14, and 15 – 16, is 11 (i.e. 2, 4, and 5, respectively). Since there are 100 observations in the data, the first three categories include 11 per cent. The last ten, of course, include the remaining 89 per cent. If the units on the abscissa represent a measure of weight, we might wish to know below what point we would find 11 per cent of the observations. Conversely, we might wish to know what proportion of the observations exceed a given weight. Figure 2.5 indicates that 11 per cent of the observations lie between 11 and 14. In Figure 2.5 we have essentially a three-dimensional polygon. This figure illustrates what is meant by obtaining a proportion of the area in a graph. Here 11 per cent of the area has been separated from the rest of the figure to further emphasize the concept. It can be seen clearly that the upper boundary of the figure, the "curve," has no particular meaning unless it forms, with the base, a closed figure with a finite area.

There is a method we may use to obtain proportions of a graphed set of data that simplifies and makes more precise the procedure for determining the limits for finite proportions. This method is employed in *cumulative* data graphs.

The Cumulative Frequency Polygon

The grouped data in Table 2.3 are used to construct the cumulative frequency polygon in Figure 2.6. The cumulative frequency distribution is shown in Table 2.4.

Interval	Frequency	Cumulative frequency
11 - 12	2	2
13 - 14	4	6
15 - 16	5	11
17 - 18	7	18
19 - 20	11	29
21 - 22	16	45
23 - 24	19	64
25 - 26	13	77
27 - 28	9	86
29 - 30	6	92
31 - 32	5	97
33 - 34	2	99
35 - 36	1	100

TABLE 2.4

To obtain the cumulative frequency, the observations in the first category are added to those in the second, those in the first and second are added to those in the third, and so on. Thus for each category there is a numerical value that represents the number of observations in all categories up to and including that particular category. In Figure 2.6 the height of the ordinate represents the cumulative frequency. The cumulative frequency polygon is not isomorphic to the frequency polygon we examined earlier, and the area relationship of the histogram and the frequency polygon, discussed above, does not apply to the cumulative frequency polygon.

Proportions and Quantiles

In addition to graphic methods for summarizing data, it is often desirable to use numerical indices which have the property of dividing the distribution into segments containing known proportions. These indices are known as

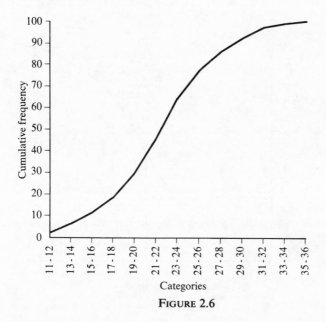

Categories

FIGURE 2.6

quantiles. Three of the more commonly used quantiles are the *percentile,* *decile,* and *quartile.*

Quantiles

The *quantile* used to evaluate the relative position of a given observation in a frequency distribution depends on the precision with which it is essential to indicate relative standing. (Figure 2.7 outlines the major quantiles.) When the units on the abscissa reflect small and critical differences, percentiles may be more nearly adequate. Quite frequently, however, abscissa units (categories) are, at best, gross estimates of differences between observations. In that case, deciles or quartiles may be indicated. Quantiles are often used as the basis for criteria in decision-making. This is particularly true when the decision involves a critical cut-off. For instance, if an employer wishes to employ one-quarter of the persons who applied for a position with his firm, he may simply determine who placed in the 4th quartile on a particular aptitude test score distribution. It is not necessary to calculate the percentile rank of each applicant.

Percentiles

Percentiles are arbitrary points which divide a frequency distribution into 100 equal quantiles. The percentile distribution is based on the number of observations constituting a given per cent of the total number of observations in the distribution, irrespective of category. Percentiles are frequently used to

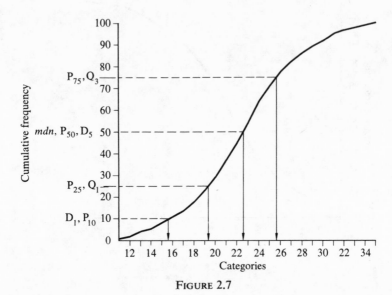

FIGURE 2.7

determine what proportion of the distribution falls below a given category. The cumulative frequency distribution may be used to determine the percentile equivalent of a given category. (Later we shall again return to percentile as an index of relative performance.)

Let us assume, for example, that the ungrouped data presented earlier consists of scores obtained by 100 children in an examination. Frequently there is no obvious relationship between test raw score and the extent to which these scores represent the performance of the children. As we have previously seen, the scores ranged from 11 to 35. We might wish to know, for instance, how the performance of a given child compares with that of others who took the same test. Test scores have only a comparative meaning. They often constitute, at best, an ordinal scale. How shall we describe the performance of an individual who received the score of 33 on the test? In itself, the score is not meaningful. If you know, however, that 97 children obtained scores lower than 33, you might gather that this level of performance on the test, relative to that of the others, is very high. When we say that an individual whose score is 33 demonstrates a high level of performance on the given test, we imply that such a score is relatively infrequent and it may be assumed that it is highly improbable that someone got such a score on the basis of chance alone—improbable, but not impossible. After all, if 97 per cent of the students who took the test can be expected to obtain lower scores, it may be inferred that, were a random sample of scores selected from the group, only 2 per cent of the time would the score 33 be exceeded in the long run.

Deciles

The tenth percentile (P_{10}) is called a *decile*—the first decile (D_1). Frequently, it is not essential to be able to discriminate between percentiles. In that case, deciles may be adequate. The 50th percentile is called the *median* (*mdn*).

The median (*mdn*) divides a distribution in two equal parts. The median is the middle observation in a distribution: 50 per cent of the distribution falls above and 50 per cent below that observation.

Deciles are multiples of the 10th percentile. In a distribution, there are 99 percentiles and 9 deciles. These units divide the distribution into 10 equal portions.

The frequency distribution may also be divided into four portions with equal frequency of observations in each portion. Each such portion is called a *quartile*.

Generally, if the abscissa units in a frequency distribution represent performance score categories for a test that is not reliable, the different categories may not discriminate adequately between individuals who obtained different scores when the differences between these categories are not very extensive. For instance, the difference between an individual who obtains an I.Q. score of 103 and one who obtains an I.Q. score of 100 cannot be considered staggering. On the other hand, an individual who can remain under water 103 seconds is doing considerably better than one who can remain submerged only 100 seconds.

Problems

Given the following 30 observations:

```
5  6  2  4  6  8  1  3  5  2
7  4  5  5  0  1  3  4  5  3
5  9  7  4  4  3  2  8  6  7
```

1. Determine the range.
2. Set up a tally-sheet frequency distribution.
3. Construct a frequency histogram.
4. Convert the histogram to a frequency polygon.
5. Construct a cumulative frequency polygon.
6. Construct a per cent cumulative frequency polygon.
7. Determine:
   a. The median.
   b. The first and third quartiles.
   c. P_{15} and P_{94}.

Answers

1. The range:

The simplest procedure is to begin by arranging the observations in order from lowest to highest:

$$
\begin{array}{cccccc}
0 & 2 & 4 & 5 & 5 & 7 \\
1 & 3 & 4 & 5 & 6 & 7 \\
1 & 3 & 4 & 5 & 6 & 8 \\
2 & 3 & 4 & 5 & 6 & 8 \\
2 & 3 & 4 & 5 & 7 & 9
\end{array}
$$

The range—$w = X_h - X_l$

$= 9 - 0$

$= 9$

2. Tally-sheet frequency distribution.

Categories	Tally	Frequency
0	|	1
1	||	2
2	|||	3
3	||||	4
4	＋＋＋	5
5	＋＋＋ |	6
6	|||	3
7	|||	3
8	||	2
9	|	1

$$\Sigma f = 30$$

3. Frequency histogram.

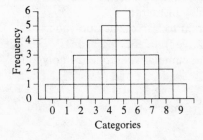

4. Frequency polygon.

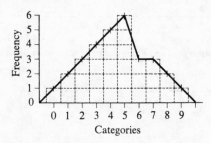

5. Cumulative frequency polygon.

Categories	Tally	Cumulative frequency
0	|	1
1	||	3
2	|||	6
3	||||	10
4	⧣⧣	15
5	⧣⧣ |	21
6	|||	24
7	|||	27
8	||	29
9	|	30

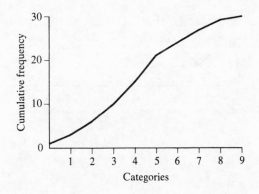

6. Per cent cumulative frequency polygon.

Category	Frequency	Per cent	Per cent cumulative frequency
0	1	3	3
1	2	7	10
2	3	10	20
3	4	13	33
4	5	17	50
5	6	20	70
6	3	10	80
7	3	10	90
8	2	7	97
9	1	3	100

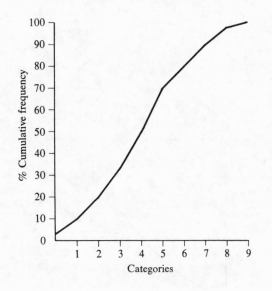

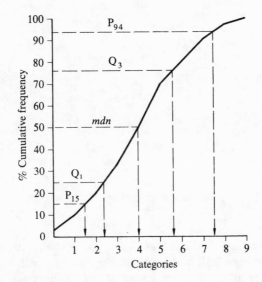

III

Elementary Probability

For the scientific investigator, the *probability* of an event E is the likelihood that the event will be observed, under specified conditions, with a determinable frequency. Generally, the event is one of several possible alternative events (E's) whose frequency of occurrence in the future is uncertain. It can be specified, then, that there are a number of possible events one or more of which will be observed. And, barring biasing factors, the probability of the event E is related to the proportion of the different alternative events (E's) that constitute the population or sample being investigated.

Determining the Probability of an Event E

We will make the assumption, in the examples that follow, that the selection of the event E is based on a process of fair selection. This means that the selection process is equally favorable to all possible events (E's) which may be selected from a given set. The selection process is *unbiased*. Dice that are unevenly constructed in weight or shape, coins that are not true, etc., result in biased event selection; they form unique probability situations which are not particularly relevant to our examination of the hypothetical and empirical estimates of the probability of an event.

Below, the term P represents probability, and P(E) the probability of the event E. The probability that E will not be observed is the probability of *not E*. The probability of *not E*, or $P(\bar{E})$, is defined as

$$P(\bar{E}) = 1.00 - P(E)$$

When more than one event is considered, we indicate each one with a subscript, i.e. E_1, E_2, E_3, E_4, etc.

34

The Simplest Case

An event (E) that can be observed to occur E times out of E's times has a proba-bility $\frac{E}{E's}$:

$$P(E) = \frac{E}{E's}$$

Example: what is the probability of obtaining a head (E) on the toss of a coin?

We are concerned with the outcome of an event E, a head. There are two possible outcomes (E's), head and tail. Thus, E = 1, and E's = 2.

$$P(E) = \frac{E}{E's}$$

$$P(E) = \tfrac{1}{2}$$

$$= .50$$

What is the probability of *not* obtaining a head? The probability of *obtaining* a head is .50.

$$P(\not E) = 1.00 - P(E)$$

$$= 1.00 - .50$$

$$= .50$$

Example: what is the probability of obtaining a four when rolling a die?

We are concerned with the outcome of the event four, out of six possible events. Thus, E = 1, and E's = 6.

$$P(E) = \frac{E}{E's}$$

$$P(4) = \tfrac{1}{6}$$

$$= .167$$

Example: what is the probability of obtaining any value *other* than a four when rolling a die?

$$P(\not E) = 1.00 - P(E)$$

$$= 1.00 - .167$$

$$= .833$$

It can be seen that the probability of any event depends on two factors that must be determined: the number of *possible* outcomes, (E's), that con-stitute all possible alternative observations, and the number of *favorable* outcomes (E). It should be noted that in the simplest case of probability estimation, and in more complex cases also, the probability of an event is determined from its frequency of occurrence in a frequency distribution in which there is, as a rule, more than one scale category (irrespective of the type of scale).

In the cases above the frequency distribution is rectangular since the frequency in each category is *one*. The frequency distribution consists of a finite number of categories, each of which contains one observation. The difference between the simple probability estimates we deal with here and the more complex ones we will discuss in later chapters has to do with the nature of the frequency distribution of possible outcomes in a chance selection situation. In the simplest case, $\frac{E}{E's}$ depends upon the selection of E from a distribution of E's in which each category in the distribution has the same frequency: 1.00. In more complex cases, the distribution of E's is such that few, if any, events have the same frequency.

It is also essential to note that the prediction of an event based on the estimation of its probability does not apply to any momentary occurrence. Probability estimation is done in terms of what may be expected *in the long run*. It is not possible to determine what will happen on any single occasion by means of the mathematical procedures and models available to the statistician. Thus, while the probability of obtaining a head when tossing a coin is .50, the probability of obtaining a tail is also .50. Consequently, since each event is equally likely, estimating the probability of obtaining a head is not applicable to what will happen on the *next* coin toss, but what will happen *in the long run*.

The probability of $\frac{E}{E's}$ remains the same for repeated observations.

If a coin is tossed once, the probability of observing a head is the same as the probability of observing heads if we tossed the coin ten times. This is a case of sampling with *replacement:* observing an occurrence does not change its frequency in the distribution.

Mutually exclusive or independent events are events that, when selected for observation, do not result in the selection of any other events from among all possible events in the distribution. For instance, the selection of a card from a deck is an independent event. No other event is affected by this selection. There are some distributions which are not independent-event distributions. If we are concerned with the relationship between the height and weight of a population of persons, we might select a particular individual and measure his height. The corresponding observation of weight is determined since there is only one value for weight which a given individual may contribute to the population. Consequently, specifying the individual, and noting one observation, results in a non-chance selection of the corresponding observation. This kind of biasing is called *correlation*. There are other kinds of dependence.

If E_1 and E_2 are mutually exclusive events (independent events) the probability that either E_1 or E_2 will be observed is the sum of their respective probabilities.

$$P(E_1 \text{ or } E_2) = P(E_1) + P(E_2)$$

Example: what is the probability of selecting either the ace of diamonds or the ten of clubs from a standard deck of playing cards?

There are 52 cards in a deck. There is one ace of diamonds and one ten of clubs.

The probability of selecting an ace of diamonds is

$$P(E_1) = \frac{E_1}{E's}$$

$$P(\text{ace of diamonds}) = \frac{1}{52}$$

$$= .019$$

Likewise, the probability of selecting a ten of clubs is

$$P(E_2) = \frac{E_2}{E's}$$

$$P(\text{ten of clubs}) = \frac{1}{52}$$

$$= .019$$

Then the probability of selecting either the ace of diamonds *or* the ten of clubs is

$$P(E_1 \text{ or } E_2) = P(E_1) + P(E_2)$$

$$= .019 + .019$$

$$= .038$$

Example: what is the probability of obtaining a four *or* a six when throwing a die?

The probability of a four is

$$P(E_1) = \frac{E_1}{E's}$$

$$P(\text{four}) = \tfrac{1}{6}$$

$$= .167$$

The probability of a six is

$$P(E_2) = \frac{E_2}{E's}$$
$$P(six) = \tfrac{1}{6}$$
$$= .167$$

The probability of a four *or* a six is

$$P(E_1 \text{ or } E_2) = P(E_1) + P(E_2)$$
$$= .167 + .167$$
$$= .334$$

If E_1 and E_2 are mutually exclusive events, the probability that E_1 and E_2 will both be observed is the product of their respective probabilities.

$$P(E_1 \text{ and } E_2) = P(E_1) P(E_2)$$

Example: when tossing two coins, what is the probability of observing two heads?

It was previously shown that the probability of obtaining a head, on the toss of a coin is .50. Thus, the probability that the outcome of the toss of two coins will be head, head, is

$$P(E_1 \text{ and } E_2) = P(E_1) P(E_2)$$
$$= (.50)(.50)$$
$$= .25$$

Example: when throwing two dice, what is the probability of observing the value four?

The value four may be obtained by the following combination of surfaces of the dice:

	die 1	die 2
E_1	1	3
E_2	2	2
E_3	3	1

The probability of obtaining the combination 1 and 3, E_1, is (.167)(.167), or .028.

The probability of obtaining the combination 2 and 2, E_2, is also .028.

Finally, the probability of E_3, the combination 3 and 1 is again .028.

The probability that the value four will be observed when the dice are tossed depends upon any one of the three events above occurring. That probability is

$$P(E_1 \text{ or } E_2 \text{ or } E_3) = P(E_1) + P(E_2) + P(E_3)$$
$$= (.028) + (.028) + (.028)$$
$$= .084$$

In this example we have combined the *additive* and *multiplicative* rules for determining the probability of a particular event.

Conditional Probability

Under certain circumstances, when sampling without replacement, for instance, the probability of events depends on other events that preceded the specified observation.

If the events E_1 and E_2 are *not* mutually exclusive, e.g. not independent, their probability is the product of their respective conditional probabilities.

Permutations

A *permutation* is the arrangement of objects in a set. For instance, given the letters A, B, and C, there are six possible permutations of these letters:

$$\text{ABC} \quad \text{BCA} \quad \text{CAB}$$
$$\text{ACB} \quad \text{BAC} \quad \text{CBA}$$

When there is a relatively small number of objects in the set—for instance, 2, 3, or 4—it is relatively simple to determine the number of permutations by actually rearranging the objects. However, when there is a large number of objects, the problem becomes considerably more time-consuming and confusing. In general, to determine the number of possible permutations the procedure below may be used.

The number of permutations (P^n) of n objects is n!

The term n! is read n factorial, and is defined as

$$n! = n \cdot (n - 1) \cdot [(n - 1) - 1] \cdot \{[(n - 1) - 1] - 1\} \ldots$$

Example: how many permutations are there of the letters ABCD?

$$P^n = n!$$
$$P^4 = 4!$$
$$= 4 \cdot 3 \cdot 2 \cdot 1$$
$$= 24$$

Example: in how many different ways can eight playing cards be arranged?

$$P^n = n!$$
$$P^8 = 8!$$
$$= 8 \cdot 7 \cdot 6 \cdot 5 \cdot 4 \cdot 3 \cdot 2 \cdot 1$$
$$= 40320$$

The number of permutations of n objects, taken r at a time, is $\dfrac{n!}{(n-r)!}$.

Thus,

$$P^n_r = \frac{n!}{(n-r)!}$$

Example: how many different groups of three cards can be obtained from five different playing cards?

$$P^n_r = \frac{n!}{(n-r)!}$$

$$P^5_3 = \frac{5!}{(5-3)!}$$

$$= \frac{5 \cdot 4 \cdot 3 \cdot 2 \cdot 1}{2 \cdot 1}$$

$$= \frac{5 \cdot 4 \cdot 3 \cdot \cancel{2 \cdot 1}}{\cancel{2 \cdot 1}}$$

$$= 5 \cdot 4 \cdot 3$$

$$= 60$$

Example: given ten blocks, each of a different color, how many groups of five blocks can be obtained without repeating any color in any one group?

$$P^{10}_5 = \frac{10 \cdot 9 \cdot 8 \cdot 7 \cdot 6 \cdot \cancel{5 \cdot 4 \cdot 3 \cdot 2 \cdot 1}}{\cancel{5 \cdot 4 \cdot 3 \cdot 2 \cdot 1}}$$

$$= 10 \cdot 9 \cdot 8 \cdot 7 \cdot 6$$

$$= 30,240$$

Example: how many permutations of two objects can be obtained from four different objects?

$$P^n_r = \frac{n!}{(n-r)!}$$

$$P^4_2 = \frac{4 \cdot 3 \cdot \cancel{2 \cdot 1}}{\cancel{2 \cdot 1}}$$

$$= 12$$

If there are n objects, with r_1 alike, r_2 alike, and r_n alike, they can be arranged to obtain $\dfrac{n!}{(r_1!)(r_2!)(r_n!)}$ *permutations.*

Example: how many different permutations can be obtained from the word *committee* so that each one is distinct from the others?

There are nine letters in the word *committee*. Three single letters, c, o, i, and three sets of two letters, mm, tt, and ee. The number of permutations is

$$\frac{9 \cdot 8 \cdot 7 \cdot 6 \cdot 5 \cdot 4 \cdot 3 \cdot 2 \cdot 1}{(1)(1)(1)(2 \cdot 1)(2 \cdot 1)(2 \cdot 1)} = \frac{362880}{4} = 60{,}480$$

Example: how many different permutations can be obtained from the word *Ohio*?

$$\frac{4 \cdot 3 \cdot 2 \cdot 1}{(1)(1)(2 \cdot 1)} = 12$$

Combinations

Three coins can be arranged to show the following:

HHH, TTT, HHT, THH, HTT, THT, HTH, TTH

However, only the HHH and TTT groups are unique. The others differ in order but contain either one or two H's or T's:

HHT, THH, HTH, TTH, HTT, THT

There are three categories of unique groups: three H's or T's, two H's or T's, one H or T. When we are not particularly concerned with the order in which the H's and T's appear in the groups, but only how many there are, we determine the number of *combinations* of r objects in a set of n objects.

If there are r objects in a set of n objects, the number of combinations, ignoring the order of arrangement of the r objects, is $\dfrac{n!}{r!(n-r)!}$.

Example: how many combinations can be obtained from the letters XYZ, giving distinct groups of two letters?

$$C_r^n = \frac{n!}{r!(n-r)!}$$

$$C_2^3 = \frac{3 \cdot 2 \cdot 1}{(2 \cdot 1)(1)} = 3$$

Example: how many combinations of three cards can be obtained from a set of six different cards?

$$C_r^n = \frac{n!}{r!(n-r)!}$$

$$C_3^6 = \frac{6\cdot5\cdot4\cdot3\cdot2\cdot1}{(3\cdot2\cdot1)(3\cdot2\cdot1)}$$

$$= \frac{720}{36}$$

$$= 20$$

Discrete Probability Distributions

In previous sections we considered isolated events and examined their respective probability. Often, however, we will be concerned with the probability of each of the events in a set, or distribution of observations, e.g. E's, from which individual cases have been selected.

It may be particularly important to know the probability of observing each category in the distribution when the frequency of observations differs from category to category. The probability of selecting any one card (1/52) from a standard deck is the same as the probability of selecting any other card. This is true because each category has a frequency of 1.00. However, let us suppose that we wish to know the probability of each possible value one might obtain when two dice are thrown. It is quite clear that if we wish to know the probability of obtaining the value two, i.e. two dots, when throwing the dice, we need only know that the probability of obtaining the value one when throwing one die is 1/6, and that the same holds for the other die. The combined probability is then 1/36. It is possible to plot the probability of obtaining any total number of dots, or any value, when throwing the two dice. The frequency distribution of this is shown in Figure 3.1.

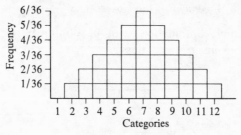

FIGURE 3.1

Note that the sum of the frequencies in all categories is equal to the denominator of the probability ratio, i.e. 36. This represents the number of possible alternative outcomes, E's.

The distribution in Figure 3.1 is a discrete variable distribution. The categories are limited to finite, discrete values. In the case above, there are twelve categories. The distribution is symmetrical, and the value we might expect most frequently is seven. The values two and twelve are to an equal degree least frequent.

The Binomial Distribution

Let us again examine the coin-tossing problem. There are certain kinds of events for which there are only two alternatives: they occur in one state or another, e.g. head *or* tail, ON *or* OFF, etc. *Binomial distributions* describe such events.

We have up to now spoken of events that take place (E), events that do *not* take place (E̶), and the probability of the occurrence of events [P(E)]. To the end of this chapter, because of the frequency with which it appears in other texts, the letter p shall represent E, and the letter q shall represent E̶, or $1.00 - p$. All possible, i.e. alternative, events are represented by p+q.

An event that has the probability P on repeated independent observations has a probability E_r of occurring r times in the n independent observations.

$$P(E_r) = C_r^n (p^r)(q^{n-r})$$

(You will recognize C_r^n as a *combination*.)
Therefore:

$$P(E_r) = \frac{n!}{r!(n-r)!} (p^r)(q^{n-r})$$

Example: a student is required to perform a finger-maze task. He has one chance in two of making the correct turn at each of ten choice points. What is the probability that he will choose seven correct turns on the first trial?

$$P(E_r) = \frac{n!}{r!(n-r)!} (p^r)(q^{n-r})$$

$$= C_7^{10}(p^7)(q^3)$$

$$= \frac{10 \cdot 9 \cdot 8 \cdot 7 \cdot 6 \cdot 5 \cdot 4 \cdot 3 \cdot 2 \cdot 1}{7 \cdot 6 \cdot 5 \cdot 4 \cdot 3 \cdot 2 \cdot 1(3 \cdot 2 \cdot 1)} (.5^7)(.5^3)$$

$$= \frac{10 \cdot 9 \cdot 8}{3 \cdot 2 \cdot 1} (.008)(.125)$$

$$= \frac{720}{6} (.001)$$

$$= .12$$

The distribution generated by all possible values of E_r, i.e. $C_r^n(p^r)(q^{n-r})$, is the binomial *distribution*. C_r^n is the binomial *coefficient*. The binomial distribution defines the probability of events with fixed category values in discrete distributions.

The binomial term $(p + q)^n$ defines the probability of each of the n events.

$$(p + q)^n = p^n + C_1^n(p)(q^{n-1}) + C_2^n(p^2)(q^{n-2}) \ldots + C_r^n(p^r)(q^{n-r}) \ldots + q^n$$

The expansion of the binomial term $(p + q)^n$ can be used to determine the probability of all events in the distribution, and, consequently, the frequency of the observations in each category.

Problems

1. a. What is the probability of obtaining four heads when eight coins are tossed?
   b. What is the probability of selecting any diamond from a standard deck of playing cards?
   c. What is the probability of obtaining a six when rolling a die?
   d. What is the probability of selecting any king from a standard deck?

2. a. What is the probability of obtaining, from a standard deck, any card other than the jack of spades?
   b. What is the probability of obtaining any value other than three when rolling a die?
   c. What is the probability of obtaining, from a standard deck, any card other than a jack, king or ace?
   d. What is the probability of not obtaining heads when two coins are tossed?

3. a. What is the probability of selecting either a diamond or a jack from a standard deck?
   b. What is the probability of selecting either a ten or a queen from a standard deck?
   c. What is the probability of obtaining either a four or a one when rolling a die?
   d. If you have five quarters, four dimes, three nickels, and three pennies in a jar, and you remove one coin at random from the jar, what is the probability that the coin is either a quarter or a penny?

4. a. Given the information in problem 3d, if you remove *two* coins, what is the probability that you will obtain a quarter and a dime?

b. What is the probability of selecting two cards from a standard deck such that the cards are the ten of diamonds and the ace of hearts?

c. When rolling two dice, what is the probability of obtaining a three and a six?

d. There are ten balls in a jar. Three are white, seven are black. What is the probability that two balls, removed at random, are both white?

5. a. You have dealt cards to each of five players, in a poker game. For each player, except the last, two cards are face down, and three cards are showing. Of the cards that are showing, not one is an ace. What is the probability that the next card you will deal is an ace?

b. Given the information in problem 5a, what is the probability that the next card you will deal is the ace of spades?

6. a. In how many different ways can the letters ABCDE be arranged?

b. How many different permutations of the letters ABCDE can be made, taken three at a time?

c. How many four-letter permutations can be made from the word *letter*?

d. How many different color groups can be obtained from a set of six different colors?

7. a. How many combinations can be obtained of six objects taken two at a time?

b. How many combinations can be obtained with the five letters ABCDE, taken four at a time?

8. a. What value would be most *likely* to be observed if two dice were thrown?

b. What value would be most *unlikely* to be observed if two dice were thrown?

9. a. A rat is running a ten-unit multiple T-maze. What is the probability that he will make three correct turns on the first trial?

b. What is the probability of obtaining three sixes if a die is thrown five times?

Answers

1. a. .50
   b. $13/52 = .25$
   c. $1/6 = .17$
   d. $4/52 = .08$

2. a. $1 - 1/52 = .98$
   b. $1 - 1/6 = .83$
   c. $1 - 12/52 = .77$
   d. $(.5)(.5) = .25$

3. a. $13/52 + 4/52 = .33$
   b. $4/52 + 4/52 = .16$
   c. $1/6 + 1/6 = .34$
   d. $5/15 + 3/15 = .53$

4. a. $5/15(5/15) = .11$
   b. $1/52(1/52) = .0004$
   c. $1/6(1/6) = .029$
   d. $3/10(3/10) = .09$

5. a. 14 cards are not showing; thus $4/38 = .11$
   b. .03, that is, $1/38$.

6. a. $5! = 120$
   b. $5!/2! = 60$
   c. $6!/1!(2!)(2!)(1!) = 180$
   d. $6! = 720$

7. a. $6!/2!(4!) = 15$
   b. $5!/4!(1!) = 5$

8. a. Seven. (See Figure 3.1.)
   b. Either two or twelve. (See Figure 3.1.)

9. a. $[10!/3!(7!)](.5^3)(.5^7) = .12$
   b. $(1/6)^3 = .005$

IV

Measures of Central Tendency

In Chapter III we reviewed some mathematical models and procedures employed in describing the probability of certain events. The scientific investigator, however, is concerned with the laws that govern the empirical world. To understand these laws, he observes the world and attempts to determine what these laws are through the description of his observations. Since the world consists of variable events, he must attempt to determine whether, in the observations he makes, there is any pattern that can be made to translate variability into order. For, if the variability of the world were the result of random events, it would be impossible to establish a discipline of science. Fortunately, there are such patterns.

One of the necessary components of science is the order that may be found in the observed variability. It is difficult for the novice to realize that order and variability are not paradoxical concepts, and that they do not form mutually exclusive categories. He should remember that variability refers to the fact that events may be assigned to different categories, while within categories, observations may be quite stable.

The investigator makes decisions about the nature of the category system (scale) he will use to classify events he is observing. Once the scale has been developed, it is essential that he determine whether there are *trends* relating categories to each other. We know, for instance, that if we hold an object and release it, it will fall to the ground. (Though this case, too, is a matter of probability—we state it otherwise because the probability that the object will *not* fall is of such magnitude that only theoretical physicists worry about it.) We are dealing here with two categories for classifying events: *yes* and *no*. That is, the event is observed to occur or not occur. In many other cases, the categories describe properties of events which may assume

47

more than two values. The observations of height, weight, intelligence, for instance assume values belonging in a large number of categories. Since it is not possible, in these cases, to describe events in terms of simple dichotomies, it is essential to determine whether there is some pattern to the frequencies with which we make observations in the various categories of the variable in question.

The patterns we are looking for are essentially descriptions of the data that will lead to a way of identifying the values of the variable which are most likely to be observed, those which are least likely to be observed, and the way in which the categories differ from each other with respect to frequency. This kind of analysis makes possible the statement of empirical laws which summarize, for a given variable, the frequency distribution. In itself, the frequency distribution is a legitimate description of the variable. It is often, however, a cumbersome affair.

Effective communication of all the information in a frequency distribution requires a great deal of effort. Usually it is done in graph form. It is therefore useful to develop a meaningful shorthand for communicating the information contained in a large number of observations. This shorthand must be based, somehow, on the more salient features of the data to be described.

It has been determined empirically that, in many cases, distributions of observations made on variable events tend to show a concentration in some categories and not in others. Empirical data suggest that the world shuns extremes. This feature of a distribution is not an essential requirement—there are many exceptions—but distributions in which there are equally distributed categories are relatively rare.

In a graph, the concentration of observations in a particular category in a distribution produces a peak for that category. This concentration, or peak, is called the *mode* (*mo*).

The Mode

The *mode* (*mo*) is the category in the distribution that contains observations with the greatest frequency. In grouped data, the mode is associated with the midpoint of the category which has the greatest frequency.

Empirically, the category with greatest frequency concentration often tends to be centrally located in a distribution. However, this need not be the case. Thus, as a measure of central tendency, the mode leaves much to be desired. Its usefulness as a measure of central tendency is artifactual to the distribution. In no way is the mode related to the center of the distribution mathematically. In addition, there are a number of computational artifacts associated with the nature of the data and the data processing procedures which tend to affect the position of the mode. For instance, grouping observations can change a distribution with many modes (*multimodal*) to a distribution of grouped data, which is *unimodal*.

To illustrate the concepts and computational procedures in this chapter, 100 observations have been obtained at random. These observations are presented in Table 4.1 as raw data. They are presented in a frequency distribution in Table 4.2, and plotted in histogram form in Figure 4.1. Table 4.3 presents the data grouped with an interval size of 4, and this grouped data appear in Figure 4.2, in histogram form.

30	7	32	14	26
26	14	29	24	9
40	32	11	31	20
36	21	24	16	20
39	11	22	14	41
34	27	30	23	38
37	27	31	22	24
10	24	24	15	16
18	26	28	31	27
39	14	17	5	23
21	33	24	26	20
12	29	9	37	26
15	20	22	14	30
25	28	21	29	23
39	36	20	23	31
8	12	20	28	21
24	22	15	28	32
12	27	20	32	16
36	12	33	22	19
28	38	27	43	29

TABLE 4.1

The frequency distribution, Figure 4.1, is decidedly multimodal. There are several distinct categories which would have produced sharp peaks were this histogram converted to a frequency polygon. The frequency distribution for *grouped* data looks altogether different. Some of the peaks have been eliminated. Thus, grouping produces a factitious change in the frequency distribution form.

There is also an interesting difference between the Figure 4.1 and Figure 4.2 graphs with respect to central tendency. In Figure 4.1, the categories 20 and

Category	Frequency		Category	Frequency
5	1		25	1
6	0		26	5
7	1		27	5
8	1		28	5
9	2		29	4
10	1		30	3
11	2		31	4
12	4		32	4
13	0		33	2
14	5		34	1
15	3		35	0
16	3		36	3
17	1		37	2
18	1		38	2
19	1		39	3
20	7		40	1
21	4		41	1
22	5		42	0
23	4		43	1
24	7			

TABLE 4.2

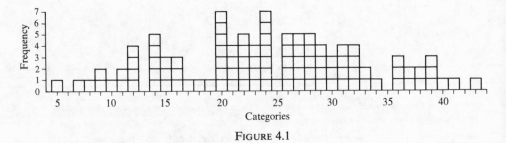

FIGURE 4.1

24 have equal numbers of observations approximately in the center of the
distribution. In Figure 4.2, grouping has combined these categories giving
the mode at the interval 21 – 24 an exaggerated concentration (the midpoint is 22.5).

The investigator would be misled about the distribution were he to use the
mode exclusively as a measure of central tendency, particularly if the mode is

Interval	Frequency
5 - 8	3
9 - 12	9
13 - 16	11
17 - 20	10
21 - 24	20
25 - 28	16
29 - 32	15
33 - 36	6
37 - 40	8
41 - 44	2

TABLE 4.3

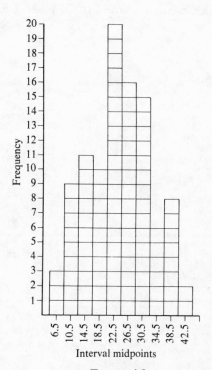

FIGURE 4.2

in a grouped frequency distribution. Data reduction always tends to produce distortion and data loss: in the raw data frequency distribution there is no observation in the category 13 or in the category 35; yet, in the grouped data, these empty categories are lost.

The simplification of frequency distributions for greater ease of handling must be considered in the context of tolerance of information loss. There are situations when considerable distortion can be tolerated. There are also situations when data cannot be reduced at all: the integrity of each observation must be maintained.

The Median

The *median* (*mdn*) is the midpoint of the frequency distribution. Half the observations in the distribution are above the median and half are below.

Computing the Median from Ungrouped Data

When the sample or population consists of an even number of observations, the median lies halfway between the two middle observations. Table 4.4 presents a sample which consists of the first 20 observations in Table 4.1.

40	36
39	33
39	32
39	29
37	28
36	27
36	27
34	27
30	26
28	24
26	22
25	21
24	20
21	14
18	14
15	12
12	12
12	11
10	7
8	

TABLE 4.4	TABLE 4.5

The observations have been arranged in ascending order. The two middle values are 26 and 28. The midpoint between them is the median, 27.

When the data consists of an odd number of observations, the median is the middle observation. Table 4.5 presents observation 21 through 39 of the data in Table 4.1. This sample has 19 observations. The middle one is 24. The median is 24.

Computing the Median from Grouped Observations

When categories are reduced by grouping, the median is determined as follows: the cumulative frequency in the grouped data frequency distribution tells us that the median must lie in the interval 21 - 24 (whose actual limits you will recall are 20.5 and 24.5). The cumulative frequency in the interval below, i.e. 17 - 20, is 33. In the interval in which the median is found, there are several observations: twenty, to be exact. Remembering that 50 per cent of the area of a histogram must be on either side of the median in a frequency polygon, we must find the value that would correspond to the 50th percentile were a cumulative frequency polygon constructed, using the *grouped* data. This is done by following the procedure below.

$$mdn = L + \left(\frac{\frac{n}{2} - \Sigma f_u}{f_i} \right) i$$

Where: L is the lower limit of the interval in which the *mdn* falls
n is the number of observations in the data
Σf_u is the number of observations below the interval in which the *mdn* falls
f_i is the number of observations in the interval in which the *mdn* falls
i is the size of the class interval

Substituting the data in Table 4.3, we obtain

$$mdn = 20.5 + \left(\frac{\frac{100}{2} - 33}{20} \right) 4$$

$$= 20.5 + \left(\frac{17}{20} \right) 4$$

$$= 20.5 + (.85)\, 4$$

$$= 23.9$$

The essential feature of measures of central tendency, and their usefulness, is that they provide indices of similarity and dissimilarity of the observations in the distribution. Let us suppose that we have a distribution consisting of the same value observed twenty times. To summarize this distribution and communicate it to others, it would be sufficient to give the value of one observation and state that it is identical to the others.

In the empirical world, however, observations in a distribution are seldom identical—the world is variable. Consequently, if there are categories which, by virture of their relatively large frequency, are most *representative* of the distribution, we might simply indicate those categories, pointing out in what way the remaining observations, of which there are relatively few, differ in frequency from the most representative ones. The *mdn* and the *mo* are not adequate to fulfill the criterion of indicating central tendency; they depend too much on area relationships and too little on the quantitative relationship of the values of the variable. The *mo* (mode) is the most frequent category and the *mdn* (median) is an arbitrary point which divides the area of a frequency polygon.

The Mean

The *mean* ($\bar{X}$, read X-bar) of a distribution is defined as the sum of the numerical values of the variable in the distribution, divided by the number of observations in that distribution. When the distribution is a sample:

$$\bar{X} = \frac{\Sigma X}{n}$$

When the distribution is a population, the *mean* is represented by the Greek letter μ (mu). (You will recall from our earlier discussion that *population* parametric measures are represented by Greek letters in many cases, and are referred to as *parameters*, while *sample* measures are represented by ordinary English letters and are referred to as *statistics*.)

The mean of a population is defined as

$$\mu = \frac{\Sigma X}{N}$$

The mean is the *average* value in the distribution. It indicates the most representative category in the distribution. However, the term "most representative" is not entirely clear. In order to explore that concept, let us examine a sample of 10 observations, shown in Table 4.6.

X
30
26
40
36
39
34
37
10
18
40
_____
$$\Sigma X = 310$$

$$\bar{X} = \frac{\Sigma X}{n}$$

$$= \frac{310}{10}$$

$$= 31$$

TABLE 4.6

The sum of the distribution is 310, and $\bar{X}$ is 31.0. Each observation contributes its value, and its distance from $\bar{X}$, to the mean of the distribution. Thus, the mean can be considered a mathematical fulcrum about which the distribution is balanced. This concept is illustrated in Figure 4.3.

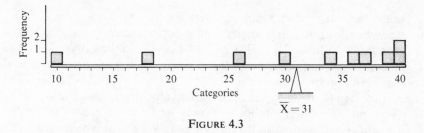

FIGURE 4.3

If you were stacking blocks whose weight corresponded to the scale values of the observations represented, then the plank on which you were stacking them would be balanced about the weight 31. Since the mean is located at the arithmetic center of the distribution, values of the variable *less* than the mean and values of the variable *greater* than the mean must be equal with respect to a weight/distance relationship. If this were not the case, no balance would ensue.

It should be pointed out, in passing, that given the mean of the distribution, and the number of observations in that distribution (n), all but one observation may be assigned at random to the distribution. However, in order that the distribution be balanced about the mean, the last observation must assume a value such that the sum of the values divided by the number of values is equal to the mean. Thus, in a distribution of this kind, there are n − 1 *degrees of freedom* (df). We will examine the concept of df in greater detail later. Suffice it to say, at this point, that n − 1 df means that all but one observation in a distribution may assume any value when the mean of the distribution is known; the last observation is fixed by the requirement that its value must be such that the distribution is balanced about the mean. This concept of df is illustrated in Figure 4.4. Nine observations have been selected at random. If the mean of the distribution is 25, the missing tenth observation can only be 40.

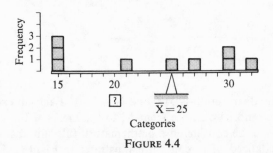

FIGURE 4.4

On the basis of random sampling, there are n − 1 ways a given distribution may be constituted and still have an expected mean.

From the definition of the mean as a fulcrum, it can be gathered that the mean derives its usefulness not only in terms of its indication of central tendency but also by virtue of the fact that the mean also takes into account the relative weight and distance of the other values in the distribution. It is, therefore, not an arbitrary measure of concentration; it is related to the dispersion of the observations in the distribution. We can also define the mean in terms of the extent to which other values in the distribution differ from it, as compared with how much they are like it.

In Table 4.7, the distribution of X has undergone a number of computational steps. These are: column 2, a linear transformation consisting of the subtraction of the mean from each observation in the distribution; column 3, the squaring of each value in column 2; column 4, the subtraction of the *mdn* (35) from each observation in the original distribution; column 5, the squaring of the values in column 4; column 6, the subtraction of an arbitrary value (18) from each observation in the distribution; column 7, the squaring of all values in column 6.

In order to understand why we have performed these computations, let us examine the following case. Suppose that the average weight of students in your class is 130 lbs. One student weighs 135 lbs. That is, that student *deviates* from the average: he *differs*. The concept of deviation is always associated with departure from some norm, usually an average or mean. Thus, the student who weighs 135 lbs deviates from the mean by 5 lbs.

Now let us return to the distribution. Column 1 consists of the original distribution of observations. This column is labeled X. The sum of the observations is 310, and the mean is 31, as we have seen previously. In column

1	2	3	4	5	6	7
X	$X - \bar{X}$	$(X - \bar{X})^2$	$X - mdn$	$(X - mdn)^2$	$X - X_9$*	$(X - X_9)^2$
30	−1	1	−5	25	+12	144
26	−5	25	−9	81	+8	64
40	+9	81	+5	25	+22	484
36	+5	25	+1	1	+18	324
39	+8	64	+4	16	+21	441
34	+3	9	−1	1	+16	256
37	+6	36	+2	4	+19	361
10	−21	441	−25	625	−8	64
18	−13	169	−17	289	0	0
40	+9	81	+5	25	+22	484
$\Sigma = 310$	$\Sigma = 0$	$\Sigma = 932$	$\Sigma = -40$	$\Sigma = 1092$	$\Sigma = 130$	$\Sigma = 2622$

* X_9 is the ninth observation; its value is 18.

TABLE 4.7

2, the mean, 31, has been subtracted from the observations. The values in column 2 are called *deviations:* the quantities by which each observation deviates from the mean is given in that column. Note that the *sum of the deviations* is zero. If we wish to get an idea of how much the observations deviate from the mean, we might take the sum of these deviations. But since that sum is zero, we cannot deal with it. Consequently, to get rid of the zero sum, we square the individual deviations (column 3) and take the *sum-of-the-squared-deviations*. Because this is a useful term, we refer to it as the *sum-of-squares*, for short. Here the sum-of-squares is 932.

In column 4, we subtract the median from each observation. The sum of the deviations of the observations about the median is not zero. This indicates that the median is not necessarily in the center of the distribution mathematically. In column 5, by squaring the values in column 4, we obtain the sum-of-the-squared-deviations about the median. This value is 1092.

Finally, in column 6, we subtract an arbitrary value, the value of the 9th observation, i.e. 18, from the observations in the distribution. Again, the sum of the deviations is not zero. By squaring these deviations about an arbitrary value in the distribution we get column 7, whose sum is 2622.

If we wished an average, or mean, extent to which the observations in the distribution deviate from each of the values we have chosen in Table 4.7, we have only to divide the sum of squared deviations by the sample size, n. This results in the *mean squared deviations*. The columns in question are 3, 5, and 7, and the sums are the sum of squared deviations about the mean (*sum-of-squares*), $\Sigma(X - \bar{X})^2$, the sum of squared deviations about the median, $\Sigma(X - mdn)^2$, and the sum of squared deviations about an arbitrary value, $\Sigma(X - X_9)^2$. Substituting the values in the table, we may compute the mean squared deviations:

$$\frac{\Sigma(X - \bar{X})^2}{n} = \frac{932}{10} = 93.2 \tag{1}$$

$$\frac{\Sigma(X - mdn)^2}{n} = \frac{1092}{10} = 109.2 \tag{2}$$

$$\frac{\Sigma(X - X_9)^2}{n} = \frac{2622}{10} = 262.2 \tag{3}$$

Note that the mean squared deviations, $\dfrac{\Sigma(X - \bar{X})^2}{n}$, in equation (1), is the smallest mean squared deviations term obtained.

It is an important property of the mean of a distribution, and in a sense a definition, that the average squared deviations (of the observations in the distribution) from the mean of the distribution is always smaller than the average squared deviations (of the observations in the distribution) from any other value in the distribution.

Thus we can see that the mean is the most representative value in the distribution since it is the value from which the other values differ least.

The Mean Computed from Grouped Data

The illustration of the procedures for computing the mean from *grouped* data employs the data in Table 4.3. In determining the mean of a distribution of grouped observations, we need to take into account that the original categories have been altered by transformation: the original categories consisted of one-unit intervals, while the new categories consist of four-unit intervals. Consequently, we assume that the observations in the intervals are distributed in the intervals so that the midpoint of the interval represents the new category, i.e. the midpoint is considered to be equivalent to the mean of the observations in the interval.

The computational procedure for determining the mean of a grouped frequency distribution is

$$\bar{X} = \frac{\Sigma(fX_{imp})}{n}$$

The terms are shown in Table 4.8.

Interval (i = 4)	Frequency (f)	Interval Midpoint (X_{imp})	(fX_{imp})
5 - 8	3	6.5	19.5
9 - 12	9	10.5	94.5
13 - 16	11	14.5	159.5
17 - 20	10	18.5	185.0
21 - 24	20	22.5	450.0
25 - 28	16	26.5	424.0
29 - 32	15	30.5	457.5
33 - 36	6	34.5	207.0
37 - 40	8	38.5	308.0
41 - 44	2	42.5	85.0

$$\Sigma(fX_{imp}) = 2390.0$$

TABLE 4.8

Substituting the data from Table 4.8, we obtain

$$\bar{X} = \frac{2390}{100}$$

$$= 23.90$$

as compared with $\bar{X} = 24.16$ computed directly from the original observations. The difference between the two is only .96 per cent. Thus it may be concluded that computation of the mean from grouped data does not introduce much error.

Problems

Given the following 50 observations:

$$
\begin{array}{rrrrr}
17 & 37 & 9 & 35 & 11 \\
11 & 8 & 22 & 20 & 27 \\
18 & 25 & 24 & 29 & 14 \\
27 & 7 & 15 & 10 & 19 \\
6 & 47 & 16 & 22 & 32 \\
22 & 23 & 31 & 21 & 15 \\
34 & 1 & 26 & 11 & 33 \\
13 & 21 & 14 & 35 & 28 \\
21 & 25 & 16 & 1 & 26 \\
29 & 23 & 19 & 16 & 30 \\
\end{array}
$$

a. Determine the mode.
b. Determine the median.
c. Determine the mean computed from raw data.
d. Determine the median computed from grouped data (i = 5.)
e. Determine the mean computed from grouped data (i = 5.)

Answers

a. The mode.

One approach is to set up a tally-sheet frequency distribution. The distribution can best be described as multimodal, there being four categories with the frequency of three: 11, 16, 21, and 22.

b. The median.

Rank the observations in ascending order. The median is the middle value and falls in the middle of the distribution between the 25th and 26th observations. The 25th observation is the value 21, and the 26th observation is the value 21 also. Clearly, the median must fall somewhere in that space. To understand at what point the 50th percentile would be drawn in a frequency polygon using these data, we have to take into account the fact that we are dealing with a value that is not represented by any of the observations in the distribution. The median falls at a point between the 25th and 26th observation:

23rd observation: 20.00
24th observation: 21.00
25th observation: 21.00
26th observation: 21.00
27th observation: 22.00

It is not logical to suppose that the median is 21.5 simply because

$$\frac{21 + 21}{2} \neq 21.5$$

We might, however, suppose that the median must occupy some space between the value 20 and the value 22 in the distribution. We can assign an equal portion of the space between 22 and 20 to each of the values 21:

23rd observation: 20.00
24th observation: 20.50
25th observation: 21.00
26th observation: 21.50
27th observation: 22.00

The median lies between the 25th and 26th observation and is, therefore, 21.25.

c. The mean ($\bar{X}$) computed from raw data.

$$\bar{X} = \frac{\Sigma X}{n}$$

$$= \frac{1042}{50}$$

$$= 20.84$$

d. The median computed from grouped data.*

$$mdn = l + \left(\frac{\frac{n}{2} - \Sigma f_u}{f_i}\right) i$$

$$= 20.5 + \left(\frac{25 - 23}{11}\right) 5$$

$$= 20.5 + .91$$

$$= 21.41$$

* The necessary information is contained in the table on the next page.

Interval (i = 5)	f	Interval Midpoint (X_{imp})	(fX_{imp})
1 - 5	2	3	6
6 - 10	5	8	40
11 - 15	8	13	104
16 - 20	8	18	144
21 - 25	11	23	253
26 - 30	8	28	224
31 - 35	6	33	198
36 - 40	1	38	38
41 - 45	0	43	0
46 - 50	1	48	48

$$\Sigma = 1055$$

e. The mean computed from grouped observations.

$$\bar{X} = \frac{\Sigma(fX_{imp})}{n}$$

$$= \frac{1055}{50}$$

$$= 21.1$$

V

Measures of Variability

In Chapter IV we examined *measures of central tendency*. The median (*mdn*) and mean ($\bar{X}$), and, less frequently, the mode (*mo*), are used to indicate general trends in the observations. These measures are used to indicate the extent to which the observations in a distribution are similar to each other. The measures of central tendency are values which are considered to be most representative of the other observations in a distribution. The concept of *central tendency* is similar, in a way, to the social phenomenon of *stereotype*. That is, it is a composite of considerable numbers of information units. (The human central nervous system seems incapable of dealing at the level of individual differences; it appears, instead, to be organized to seek intra-individual similarities.)

The mean of the distribution, as we saw, is the most representative value in the distribution since it is that quantity from which the other values in the distribution differ least. But a property of the distribution is generally the differential frequency distribution among the scale categories. Thus, many observations may differ considerably from the mean in a distribution. The difference between each observation and the mean of the distribution is called a deviation: the extent to which observations differ from the mean of the distribution is the extent to which they deviate from the mean.

There are several indices that may be used to summarize the way in which observations in a distribution deviate from the mean of the distribution. These indices provide a measure of the data *variability*.

The Range (w) as a Measure of Variability

The *range* (w) indicates the spread of the observations in a distribution, irrespective of $\bar{X}$ or median, by specifying the number of categories in which observations were noted. A frequency distribution requiring 40 categories is more variable than a frequency distribution requiring 25 categories if the two distributions have the same number of observations. Populations and samples may vary greatly with respect to their range, and samples obtained at random from a population may also be expected to differ with respect to their range.

For instance, let us suppose that the 100 observations in Table 2.1, p. 16, represent a population. On the basis of random sampling, 25 samples of 10 observations are constituted from the population. Their range is calculated; it appears in the frequency distribution in Table 5.1(a).

Sample	range (w)	Mean ($\bar{X}$)
1	14	24.4
2	15	20.6
3	15	22.3
4	17	23.2
5	13	24.6
6	15	23.9
7	15	20.6
8	14	22.8
9	23	24.6
10	12	23.0
11	19	24.0
12	23	22.8
13	8	26.8
14	16	24.2
15	14	24.4
16	19	23.7
17	15	21.0
18	14	24.1
19	14	24.8
20	15	23.8
21	18	24.4
22	10	25.9
23	10	21.5
24	17	24.3
25	17	26.3

$$\Sigma = 592.0$$

TABLE 5.1(a)

1	2	3	4	5	6	7	8	9	10	11	12	13
19	27	26	27	30	27	27	29	26	21	20	13	23
28	17	28	26	23	19	17	21	11	30	31	23	26
22	21	21	13	29	26	21	15	24	19	14	22	30
21	17	21	20	17	22	17	21	33	27	20	11	24
29	23	13	23	29	22	23	20	17	18	24	24	30
32	12	24	23	28	21	12	25	13	20	27	24	26
31	21	20	30	24	34	21	26	34	28	16	34	24
25	23	25	22	18	22	23	18	29	21	29	20	26
19	25	20	21	23	21	25	25	31	20	26	25	28
18	20	25	28	25	25	20	28	28	26	33	32	31

14	15	16	17	18	19	20	21	22	23	24	25
15	25	14	29	26	23	24	32	31	23	20	25
27	31	31	17	32	25	20	14	21	16	26	26
21	21	31	23	19	18	29	31	29	25	23	30
31	19	33	17	28	28	21	20	31	23	31	18
23	20	22	14	20	28	19	19	21	20	15	32
21	21	18	23	23	24	25	28	25	26	19	35
28	33	24	16	18	26	34	23	24	22	32	23
24	21	25	22	21	22	20	25	29	24	24	26
24	30	23	25	25	22	23	27	27	18	21	25
28	23	16	24	29	32	23	25	21	18	32	23

TABLE 5.1(b)

The 25 sample ranges have a range of 15. Note that the range of the population is 24. Sampling ranges from a population, at random, will produce sample ranges that vary from each other. Since the observations in the samples are obtained at random, and since they vary from sample to sample, it follows that ranges that depend on the observations in the samples are likely to vary as well.

In addition to the range of each sample, the mean is also given in Table 5.1(a). Note that the means also vary from each other. The mean of a sample is a statistic. The distribution of the means of the samples is a distribution of statistics.

A distribution of statistics is called a *sampling distribution*. The distribution of the sample means is a sampling distribution of means, and the distribution of ranges is a sampling distribution of ranges. The beginner often confuses the term *distribution of the sample* (distribution of *observations*, e.g. X) with the term *sampling distribution* (distribution of *statistics*).

Note that the mean of the sampling distribution of means is 23.68, and the mean of the distribution of 250 observations computed directly is 23.68. The mean of the sampling distribution of ranges is not a particularly useful term: it cannot be equal to the population range, nor can it be a reliable index of variability since no sample range can exceed the population range and this places a limiting factor on the nature of such a distribution.

Other measures of data variability may also be used. One such measure is the *interquartile range*.

The Interquartile Range (Q)

The *interquartile range* (Q) is most often used as a measure of distribution asymmetry. That is, it indicates marked *skewness* of a distribution. When a distribution is not symmetrical about its mean, it is said to be *skewed*. Many useful distributions are not symmetrical about their mean. We will discuss the form of distributions in greater detail a little later.

Q is equal to the range of the middle 50 per cent of the observations in the distribution. It is equal to the range of the observations that lie between the 25th and 75th percentile, or Q_1 and Q_3.

Occasionally, half the interquartile range is used as the index of variability. This index is called the *semi-interquartile range*.

When the distribution is symmetrical, Q_1 and Q_3 will be equidistant from the median (Q_2). When the distribution is not symmetrical, that is, when it is *skewed*, Q_1 and Q_3 will not be equally distant from the median. When the distribution is skewed to the right, the distance $Q_3 - Q_2$ will be greater than the distance $Q_1 - Q_2$. When the distribution is skewed to the left, the distance $Q_1 - Q_2$ will be greater than the distance $Q_3 - Q_2$.

While the index Q is somewhat more precise and stable than the range as a measure of variability, it still is not sufficiently precise to be of universal value. The index Q takes into account the form of the distribution that is related to the way in which the observations are distributed among the categories, while the range takes into account only the number of categories in which there are observations. However, Q fails to distinguish between certain kinds of distributions whose variability is different.

The Variance (S^2) and Standard Deviation (S)

The *variance* (S^2) of a distribution is the mean of the squared deviations of the observations in a distribution from the mean of the distribution. We used this term previously in connection with the definition of the mean of the distribution.

The variance of a population is defined as

$$\sigma^2 = \frac{\Sigma(X - \mu)^2}{N}$$

The term σ (Greek *sigma*) represents a population *parametric value*, and μ, you will recall, is the mean of a population. N is the size of the population (number of observations).

The variance of a sample is defined as

$$S^2 = \frac{\Sigma(X - \bar{X})^2}{n}$$

It should be noted that some statisticians prefer the following procedure for the computation of the variance, particularly when the size of the sample (n) whose variance is being computed is small. For the sample statistic, S^2, the denominator of the ratio is degrees of freedom $(n - 1)$, rather than n, the sample size.

$$S^2 = \frac{\Sigma(X - \bar{X})^2}{n - 1}$$

The ratio employing df $(n - 1)$ in the denominator is called an *unbiased estimate* of the population variance (σ^2), while the ratio using n in the denominator is called a *biased estimate* of the population variance. The reason for this distinction will become clearer in later sections. It should be noted, however, that the $n - 1$ correction is not significant unless the sample size is extremely small. Since we have defined the variance as the mean squared deviations, and we are not yet concerned with estimating population parameters from sample statistics, we shall for the moment use n, rather than $n - 1$.

The *standard deviation* is the square root of the variance. The standard deviations for *population* and for *sample* are, respectively,

$$\sigma = \sqrt{\sigma^2}, \quad \text{i.e.} \quad \sigma = \sqrt{\frac{\Sigma(X - \mu)^2}{N}}$$

and

$$S = \sqrt{S^2}, \quad \text{i.e.} \quad S = \sqrt{\frac{\Sigma(X - \bar{X})^2}{n}}$$

In the computational procedures that follow, the sum of squared deviations (sum-of-squares) of the observations about the distribution mean, i.e. $\Sigma(X - \bar{X})^2$, is denoted Σx^2. The upper case roman letter X ordinarily denotes an observation. The lower case italic letter x denotes a deviation, i.e. $X - \bar{X}$. Thus, x is a symbol indicating a mathematical operation: the subtraction of the mean of the distribution from an observation X.

There are several methods for computing the variance and standard deviation of a distribution, and the computational method used will depend upon the nature of the original data. The various methods are given below.

S^2 and S Computed from Ungrouped Data

The computational procedures are based on the 20 observations in Table 5.2.

X	$(X - \bar{X})$ x	x^2		X	X^2
19	−5	25		19	361
28	+4	16		28	784
22	−2	4		22	484
21	−3	9		21	441
29	+5	25		29	841
32	+8	64		32	1024
31	+7	49		31	961
25	+1	1		25	625
19	−5	25		19	361
18	−6	36		18	324
14	−10	100		14	196
31	+7	49		31	961
31	+7	49		31	961
32	+8	64		32	1024
22	−2	4		22	484
18	−6	36		18	324
24	0	0		24	576
25	+1	1		25	625
23	−1	1		23	529
16	−8	64		16	256
$\Sigma = 480$	$\Sigma = 0$	$\Sigma = 622$		$\Sigma = 480$	$\Sigma = 12142$
	TABLE 5.2(a)			TABLE 5.2(b)	

Method 1

This is the "long" method, in which each observation is transformed to a deviation by subtraction of the distribution mean. Substituting the data in the table,

$$S^2 = \frac{\Sigma(X - \bar{X})^2}{n} = \frac{\Sigma x^2}{n} \quad \text{and} \quad S = \sqrt{S^2}$$

$$= \frac{622}{20} \qquad\qquad\qquad = \sqrt{31.1}$$

$$= 31.1 \qquad\qquad\qquad = 5.58$$

Method 2

This method involves fewer steps and is therefore simpler. However, it tends to result in relatively large sums and is particularly recommended when a calculator is available.

$$\Sigma x^2 = \Sigma X^2 - \frac{(\Sigma X)^2}{n} \qquad\qquad S^2 = \frac{\Sigma x^2}{n}$$

$$= 12{,}142 - \frac{(480)(480)}{20} \qquad\qquad = \frac{622}{20}$$

$$= 12{,}142 - 11{,}520 \qquad\qquad = 31.1$$

$$= 622$$

The computation for the standard deviation is, of course, the same as in Method 1. It is simply the square root of the variance.

S^2 and S Computed from Grouped Observations

When the data consist of large numbers of observations, it is common to group them to simplify the computation of the mean of the distribution. The variance and standard deviation can also be calculated from grouped frequency distributions. The procedure described below may be used. The data in the illustrative example, Table 5.3, are the same as those in Table 5.2(a) and 5.2(b).

$$\Sigma x^2 = \left(\Sigma f x'^2 - \frac{(\Sigma f x')^2}{\Sigma f}\right) i^2$$

$$= \left(601 - \frac{(95)(95)}{20}\right) 4$$

$$= (601 - 451.25)4$$

$$= 599$$

$$S^2 = \frac{\Sigma x^2}{n} \qquad\qquad S = \sqrt{S^2}$$

$$= \frac{599}{20} \qquad\qquad = \sqrt{29.95}$$

$$= 29.95 \qquad\qquad = 5.47$$

Interval (i = 2)	Frequency f	X_{imp}	X'	fX'	fX'^2
14 - 15	1	14.5	0	0	0
16 - 17	1	16.5	1	1	1
18 - 19	4	18.5	2	8	16
20 - 21	1	20.5	3	3	9
22 - 23	3	22.5	4	12	48
24 - 25	3	24.5	5	15	75
26 - 27	0	26.5	6	0	0
28 - 29	2	28.5	7	14	98
30 - 31	3	30.5	8	24	192
32 - 33	2	32.5	9	18	162
	$\Sigma = 20$			$\Sigma = 95$	$\Sigma = 601$

Where: X_{imp} is the interval midpoint.

X' is a transformation of X_{imp} by the subtraction of the lowest interval midpoint and the division by i $\left(\dfrac{X_{imp} - X_{imp_1}}{i}\right)$. This is done to reduce the values in the fX' and fX'^2 columns.

fX' is the product of f and X'.

fX'^2 is the product of fX' and X'.

TABLE 5.3

The standard deviation obtained when computing from a grouped frequency distribution is 5.47, while the standard deviation computed from the original observations is 5.58. The difference is not very large: it is less than 5 per cent. However, such a small difference is due at least in part to the fact that i, the size of the *class interval*, was small (2). Were the interval greater, let us say, than 10, the difference between the two differently computed standard deviations would probably have been slightly greater, though we still would not expect it to be excessive.

Transformation of a Distribution

Occasionally, a distribution may consist of relatively large values, or it may consist of negative values as well as positive ones. When the quantitative value of the observations is large, it may be convenient to reduce them by subtracting a constant value (K) from each observation in the distribution.

If some values in a distribution are negative, it may be desirable to transform the distribution so that all values are positive by adding a constant (K) equal to the highest negative value: this will transform that value to zero and the others will be positive. Other transformations may be appropriate.

In Chapter IV, we subtracted the midpoint of a category interval from each of the category midpoints and divided the remainder by the size of the category interval:

$$X' = \frac{(X_{\text{1mp}} - X_{\text{lowest}_{\text{imp}}})}{i}$$

This transformation is, in fact, two transformations: subtraction of a contant, and division by a constant. This kind of transformation is ordinarily known as *coding:* it codes category midpoints.

It might be well to examine the effects of transformation of a frequency distribution on its mean and on its standard deviation. The distribution in the illustrative example consists of 10 observations shown in Table 5.4, together with the mean and standard deviation.

X	x	x^2	
3	-2	4	
4	-1	1	$\bar{X} = \frac{\Sigma X}{n} = \frac{50}{10} = 5.0$
4	-1	1	
5	0	0	
6	$+1$	1	$S = \sqrt{\frac{\Sigma x^2}{n}} = \sqrt{\frac{44}{10}} = 2.09$
7	$+2$	4	
7	$+2$	4	
9	$+4$	16	
3	-2	4	
2	-3	9	
$\Sigma = 50$	$\Sigma = 0$	$\Sigma = 44$	

TABLE 5.4

Transformation by Adding a Constant

In the sample distribution, Table 5.4, the mean is 5.0, the standard deviation is 2.1, S^2 is 4.4, and Σx^2 is 44. In the transformed distribution—transformed by the addition of the *constant*, 2, to each observation—the new $\bar{X}$ is 7.0, Σx^2 is 44, S^2 is 4.4, and S is 2.1. Thus, transformation by addition of a constant adds the constant to the mean of the distribution but does not alter the *sum-of-squares*, variance, or standard deviation. The transformed distribution is shown in Table 5.5.

X	(X + 2)	x	x^2
3	5	−2	4
4	6	−1	1
4	6	−1	1
5	7	0	0
6	8	+1	1
7	9	+2	4
7	9	+2	4
9	11	+4	16
3	5	−2	4
2	4	−3	9
$\Sigma = 50$	$\Sigma = 70$	$\Sigma = 0$	$\Sigma = 44$

$$\bar{X} = \frac{\Sigma(X + 2)}{n} = \frac{70}{10} = 7.0$$

$$S = \sqrt{\frac{\Sigma x^2}{n}} = \sqrt{\frac{44}{10}} = 2.09^*$$

* The small differences are due to rounding off errors.

TABLE 5.5

Transformation by Subtracting a Constant

Table 5.6 presents the distribution in Table 5.4 transformed by subtraction of the constant 2. Compare this with Table 5.4. Note that the transformation subtracts the constant from the mean of the distribution, but the terms Σx^2, S^2, and S are the same in the two distributions.

X	(X − 2)	x	x^2
3	1	−2	4
4	2	−1	1
4	2	−1	1
5	3	0	0
6	4	+1	1
7	5	+2	4
7	5	+2	4
9	7	+4	16
3	1	−2	4
2	0	−3	9
$\Sigma = 50$	$\Sigma = 30$	$\Sigma = 0$	$\Sigma = 44$

$$\bar{X} = \frac{\Sigma(X - 2)}{n} = \frac{30}{10} = 3.0$$

$$S = \sqrt{\frac{\Sigma x^2}{n}} = \sqrt{\frac{44}{10}} = 2.09^*$$

* The small differences are due to rounding off errors.

TABLE 5.6

Transformation by Multiplying by a Constant

Transformation of the Table 5.4 data by multiplying by the constant 2 has the following effect: The transformed mean is *twice* the original mean, i.e.

the mean has been multiplied by the *constant*, Σx^2 has been multiplied by the *square* of the constant, S^2 has been multiplied by the *square* of the constant, and S has been multiplied by the *constant*.

X	2(X)	x	x^2
3	6	−4	16
4	8	−2	4
4	8	−2	4
5	10	0	0
6	12	+2	4
7	14	+4	16
7	14	+4	16
9	18	+8	64
3	6	−4	16
2	4	−6	36
$\Sigma = 50$	$\Sigma = 100$	$\Sigma = 0$	$\Sigma = 176$

$$\bar{X} = \frac{\Sigma[2(X)]}{n} = \frac{100}{10} = 10$$

$$S = \sqrt{\frac{\Sigma x^2}{n}} = \sqrt{\frac{176}{10}} = 4.19*$$

* The small differences are due to rounding off errors.

TABLE 5.7

Transformation by Dividing by a Constant

The transformation of the sample data in Table 5.4 by dividing by the constant 2 has this effect: $\bar{X}$ is divided by the constant, Σx^2 is divided by the square of the constant, the variance is divided by the square of the constant, and the standard deviation is divided by the constant.

X	$\left(\frac{X}{2}\right)$	x	x^2
3	1.5	−1.0	1.00
4	2.0	−0.5	0.25
4	2.0	−0.5	0.25
5	2.5	0.0	0.00
6	3.0	+0.5	0.25
7	3.5	+1.0	1.00
7	3.5	+1.0	1.00
9	4.5	+2.0	4.00
3	1.5	−1.0	1.00
2	1.0	−1.5	2.25
$\Sigma = 50$	$\Sigma = 25.0$	$\Sigma = 0$	$\Sigma = 11.00$

$$\bar{X} = \frac{\Sigma\left(\frac{X}{2}\right)}{n} = \frac{25}{10} = 2.5$$

$$S = \sqrt{\frac{\Sigma x^2}{n}} = \sqrt{\frac{11}{10}} = 1.05*$$

* The small differences are due to rounding off errors.

TABLE 5.8

Transformation of a Distribution: z

Above we saw that addition or subtraction of a constant adds or subtracts the constant from the mean of the distribution, but does not change the standard deviation. Multiplication or division by a constant multiplies or divides the mean by the constant and multiplies or divides the standard deviation by the constant.

There is an additional transformation common in the analysis of a distribution of observations. This is the z transformation. This transformation consists of dividing the deviations $(X - \bar{X})$ by the distribution standard deviation:

$$z = \frac{X - \bar{X}}{S} \quad \text{or} \quad z = \frac{x}{S}$$

The z transformation has several distinct advantages. It changes the mean of the distribution to zero and the standard deviation to 1.00. Irrespective of the distribution, transformation to z results in a distribution whose mean is zero and whose standard deviation is 1.00. This makes it possible to compare distributions which have different means and different standard deviations. For this reason, the z distribution is a standard distribution for the comparison of data whose statistics may be different; z is known as a *standard score*.

The transformation of a distribution to z is shown in Table 5.9.

X	x	x^2	$\dfrac{x}{S}$
3	−2	4	−.95
4	−1	1	−.48
4	−1	1	−.48
5	0	0	0
6	+1	1	+.48
7	+2	4	+.95
7	+2	4	+.95
9	+4	16	+1.90
3	−2	4	−.95
2	−3	9	−1.43
$\Sigma = 50$	$\Sigma = 0$	$\Sigma = 44$	$\Sigma = 0.00^*$

$$\bar{X} = \frac{\Sigma X}{n} \qquad S = \sqrt{\frac{\Sigma x^2}{n}}$$

$$= \frac{50}{10} \qquad\qquad = \sqrt{\frac{44}{10}}$$

$$= 5 \qquad\qquad\quad = 2.1$$

* Allowing for rounding off errors.

TABLE 5.9

Actually, z is a special case of S, the standard deviation of the distribution. That is, S is a measure of the extent to which the observations in the distribution deviate from the mean of the distribution, and z converts the deviation to a standard scale on the abscissa of the frequency distribution, thus making it possible to measure distances of observations in the distribution from the mean in z units rather than in S units. z is therefore a transformation of S as well.

To illustrate the use of the z transformation, let us suppose that you are concerned with a student who is referred to the guidance counselor for failure to achieve his "estimated" potential. The guidance office gives the student two tests. You are then informed that his performance on the first test was 115, and on the second, 55.

These test results seem, at face value, to be rather discrepant. That is, the student seems to be performing better on one test than on the other. You then inquire and find that both tests are supposed to measure approximately the same achievement. How is it, then, that the student performed twice as well on one test as on the other? If you inquire about the norms of the test, you are told that the mean of test 1 is 100 and the standard deviation is 15; on test 2, the mean and standard deviation are 50 and 5, respectively.

Converting the two test scores to z, we obtain:

Test 1.

$$z = \frac{(X - \bar{X})}{S}$$

$$= \frac{115 - 100}{15}$$

$$= +1.00$$

Test 2.

$$z = \frac{(X - \bar{X})}{S}$$

$$= \frac{55 - 50}{5}$$

$$= +1.00$$

Thus, irrespective of whatever else it may mean, the student performed equally well on both tests.

Problems

Given the following 50 observations:

17	37	9	35	11
11	8	22	20	27
18	25	24	29	14
27	7	15	10	19
6	47	16	22	32
22	23	31	21	15
34	1	26	11	33
13	21	14	35	28
21	25	16	1	26
29	23	19	16	30

1. Determine the variance and standard deviation using raw data method 1.
2. Determine S^2 and S using raw data method 2.
3. Determine S^2 and S with grouped data (i = 5).
4. What is the z equivalent of these observations:

      a. 6   b. 31   c. 47   d. 20.84   e. 15.76

Answers

1. The variance and standard deviation, method 1.

X	x	x^2	X^2
17	−3.84	14.75	289
11	−9.84	96.83	121
18	−2.84	8.07	324
27	+6.16	37.95	729
6	−14.84	220.23	36
22	+1.16	1.35	484
34	+13.16	173.19	1156
13	−7.84	61.47	169
21	+.16	.03	441
29	+8.16	66.59	841
37	+16.16	261.15	1369
8	−12.84	164.87	64
25	+4.16	17.31	625
7	−13.84	191.55	49
47	+26.16	684.35	2209
23	+2.16	4.67	529
1	−19.84	393.63	1
21	+.16	.03	441

X	x	x^2	X^2
25	+4.16	17.31	625
23	+2.16	4.67	529
9	−11.84	140.19	81
22	+1.16	1.35	485
24	+3.16	9.99	576
15	−5.84	34.11	225
16	−4.84	23.43	256
31	+10.16	103.23	961
26	+5.16	26.63	676
14	−6.84	46.79	196
16	−4.84	23.43	256
19	−1.84	3.39	361
35	+14.16	200.51	1225
20	−.84	.71	400
29	+8.16	66.59	841
10	−10.84	117.51	100
22	+1.16	1.35	484
21	+.16	.03	441
11	−9.84	96.83	121
35	+14.16	200.51	1225
1	−19.84	393.63	1
16	−4.84	23.43	256
11	−9.84	96.83	121
27	+6.16	37.95	729
14	−6.84	46.79	196
19	−1.84	3.39	361
32	+11.16	124.55	1024
15	−5.84	34.11	225
33	+12.16	147.87	1089
28	+7.16	51.27	784
26	+5.16	26.63	676
30	+9.16	83.91	900

$\Sigma = 1042$ $\Sigma = 4586.94$ $\Sigma = 26302$

$$\bar{X} = \frac{\Sigma X}{n} \qquad S^2 = \frac{\Sigma x^2}{n} \qquad S = \sqrt{91.74}$$

$$= \frac{1042}{50} \qquad = \frac{4586.94}{50} \qquad = 9.58$$

$$= 20.84 \qquad = 91.74$$

2. The variance and standard deviation, method 2.

$$\Sigma x^2 = \Sigma X^2 - \frac{(\Sigma X)^2}{n} \qquad S^2 = \frac{\Sigma x^2}{n} \qquad S = \sqrt{91.73}$$

$$= 26302 - \frac{(1042)^2}{50} \qquad = \frac{4586.72}{50} \qquad = 9.58$$

$$= 26302 - 21715.28 \qquad = 91.73$$

$$= 4586.72$$

3. The variance and standard deviation from grouped observations.

Interval (i = 5)	f	X′	fX′	fX′²
1 - 5	2	0	0	0
6 - 10	5	1	5	5
11 - 15	8	2	16	32
16 - 20	8	3	24	72
21 - 25	11	4	44	248
26 - 30	8	5	40	200
31 - 35	6	6	36	216
36 - 40	1	7	7	49
41 - 45	0	8	0	0
46 - 50	1	9	9	81

$$\Sigma = 50 \qquad\qquad \Sigma = 181 \quad \Sigma = 903$$

$$\Sigma x^2 = \left(\Sigma fX'^2 - \frac{(\Sigma fX')^2}{\Sigma f} \right) i^2$$

$$= \left(903 - \frac{(181)(181)}{50} \right) 25$$

$$= (903 - 655.22)25$$

$$= 247.78(25)$$

$$= 6194.50$$

$$S^2 = \frac{\Sigma x^2}{n} \qquad S = \sqrt{123.89}$$

$$= \frac{6194.50}{50} \qquad = 11.13$$

$$= 123.89$$

4. a. -1.55 b. $+1.06$ c. $+2.73$ d. 0.00 e. -0.53

VI

Frequency Distributions: II

Earlier we examined frequency distributions with respect to graphical and numerical presentation. The differential distribution of the observations among the scale categories has resulted in distribution forms variously described as *symmetrical, skewed, unimodal, bimodal,* or *multimodal.* Each of these terms conveys information about the observations in the distribution and helps the investigator in making decisions and inferences from the data.

Symmetrical and Skewed Distributions

When a distribution is symmetrical, the mean and median coincide, and half of the observations in the distribution (50 per cent of the distribution) is on either side of these two indices. The mode may also coincide, on occasion. But there are symmetrical distributions in which this is not the case. Figure 6.1 shows distributions in which the *mean* and *mdn* coincide, and *mo* does not.

Note that in distribution (a) the three indices coincide, while they do not coincide in distributions (b) and (c). The (b) and (c) distributions are also symmetrical, but (b) is *bimodal* and (c) is *multimodal.* Thus, the relationship between *mean, mdn* and *mo* provides information about the manner in which the observations are distributed among the categories. In the (a) distribution, any of the indices are equally valid as measures of central tendency. Quite obviously this is not the case in the (b) and (c) distributions. Yet the use of *mdn* or *mean* as measures of central tendency in the (b) and (c) distributions certainly is contrary to what we normally understand *central tendency* to imply: the concentration of the majority of the observations in the middle portion of the distribution.

When the distribution is not symmetrical about the *mean*, it is said to be

79

skewed. Ordinarily, the low scale-categories on the abscissa of a graph are on the left, and the higher values on the right. The distribution is said to be *positively skewed* when there is a greater proportion of observations in the lower categories than in the upper ones, i.e. when the tail points to the right. The distribution is *negatively skewed* when there is a greater proportion of

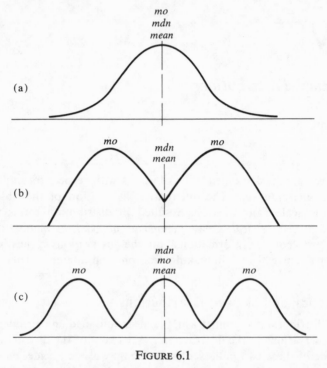

FIGURE 6.1

observations in the upper categories than there is in the lower ones, i.e. when the tail points to the left. This is illustrated in Figure 6.2. Distribution (a) is moderately skewed to the right, and (b) is moderately skewed to the left, i.e. positively and negatively, respectively. Distribution (c) is sharply skewed to the right (positive).

When a distribution is skewed, a large portion of the observations is concentrated at one end of the scale. The *mdn* will, therefore, not be in the middle of the distribution. That is, the *mdn* will not coincide with the center of the range of categories in which there are observations.

When a distribution is sharply skewed, the *mdn* will tend to be nearer the concentration of observations than the *mean,* which takes into account the distance of the furthest observation from the center of the distribution. Consequently, in sharply skewed distributions, the *mdn* is the preferred measure

of central tendency. For instance, annual salaries are reported by govern-
ment agencies in terms of *median income*. The *median* income reflects more
closely the *typical* income than the *mean* income would. This is because
most wage earners tend to receive salaries which are at the lower end of the
scale of all possible salaries. Thus, mean income would place an undue

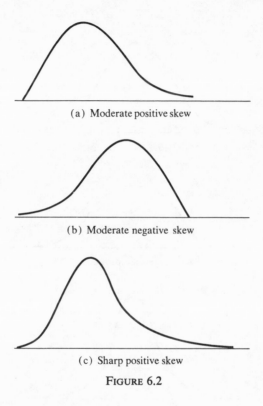

(a) Moderate positive skew

(b) Moderate negative skew

(c) Sharp positive skew

FIGURE 6.2

weight on salaries that are far up on the scale, and of which there are relatively
few.

This discussion of *skewness* of distributions serves to emphasize the departure of distributions from a useful standard distribution in psychology and
education: the *normal* distribution. We will examine this distribution in
some detail below.

Skewed distributions depart from normal in that they are not symmetrical.
However, there are also symmetrical distributions which are not normal.

The differential distribution of the observations among the scale categories
of a symmetrical distribution is known as *kurtosis*. The normal distribution

is *mesokurtic*. There are distributions which have unusually large proportions of the observations concentrated in the middle portion of the range. Such a distribution is *platykurtic*. When the observations tend to be concentrated in the *tails* of the distribution, it is said to be *leptokurtic*. These types of distributions are illustrated in Figure 6.3.

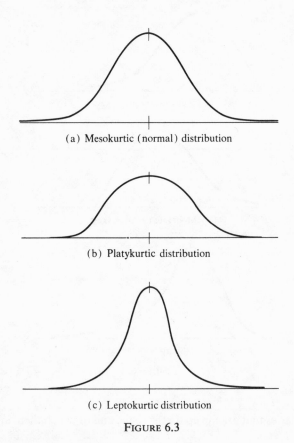

(a) Mesokurtic (normal) distribution

(b) Platykurtic distribution

(c) Leptokurtic distribution

FIGURE 6.3

The Normal Distribution

The *normal*, or *Gaussian*, distribution is a frequency distribution whose form is defined with the use of the following formula, which specifies the height of the curve at any point $X - \mu$ distant from μ:

$$Y = \frac{1}{\sigma\sqrt{2\pi}} e^{-\frac{1}{2}\left(\frac{x-\mu}{\sigma}\right)^2}$$

The term Y is an ordinate in the frequency polygon. Thus, if you know that the ordinate in the frequency polygon refers to the frequency, you can reconstruct the normal distribution from the formula. For the moment, ignore the constant $\dfrac{1}{\sigma\sqrt{2\pi}}$ and the natural logarithm e. The term $\left(\dfrac{X - \mu}{\sigma}\right)$ is familiar to you since the operation is that of the transformation z. Because the parameters σ and μ indicate that the transformation involves a population, let us refer to this transformation as Z instead of z; that is, Z is a special case of z.

To return to the formula. Y, the ordinate, indicates frequency. Thus, the frequency at any point must be a function of a constant raised to a power that is related to Z. Since Z is also a measure of the distance of a given observation, in a given category, from the mean of the distribution, the formula states, in essence, that *the height of the polygon, at any given point (ordinate) is a specifiable function of its distance from the mean.*

The normal distribution is a standard distribution in that, irrespective of the parameter plotted, if the distribution is normal, the frequency polygon will conform to the model described by the formula above. Actually, no *empirical* variables are truly normally distributed; the normal distribution is a mathematical model. In point of fact, however, there are empirical distributions that approximate the normal distribution. And, because the approximation is reasonably good, the mathematical model can be used to describe the empirical distribution that approximates it.

The normal distribution has at least one limitation that makes it impossible to fit an empirical distribution perfectly. The upper boundary of the polygon, the so-called *curve*, is *asymptotic* to the abscissa. That is, the curve touches the base only at infinity. This means, for practical purposes, that it does not touch the base. It is not possible to determine proportions of an object whose boundaries are missing. That is, it is not possible to obtain proportions of the area of a structure which is not a closed figure. Our discussion of proportions of a frequency distribution specified that the frequency polygon is a closed figure with a finite area, something like a distorted pie. For this reason alone, the normal distribution, other factors being equal, can only be an approximation of the empirical world.

In spite of this, the normal distribution is a useful and often used model of empirical distribution. Not all empirical variables are normally distributed. But such common ones as height, weight, "intelligence," and other behavior and biological and physical variables are approximately normally distributed. The relationship between the mathematical model and the life-data is established through the observation that certain empirical data appear to be distributed in the same way as the hypothetical observations in the formal model.

The normal distribution has other properties. The units on the base comprise an interval scale of units which represent equal distances from the mean (μ) of the distribution. Units representing equal distances from the mean, for instance, +1.00 and −1.00, contain equal proportions of the distribution. These proportions are fixed and known, and do not depend upon what variable is involved. Some of these proportions are shown in Figure 6.4.

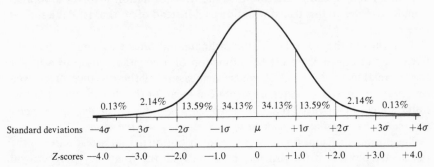

FIGURE 6.4. Proportions of the area in the normal distribution between standard deviation and Z units.

Additional proportions of the normal distribution are listed in Table A.1, p. 263. The distance X − μ, which specifies a finite relationship between a given observation and the mean of the distribution, can be converted to Z:

$$Z = \left(\frac{X - \mu}{\sigma}\right)$$

Table A.1 lists the proportion of the area between Z values and μ. Note that the height of Y, the frequency of the category in which X is included, is also given. The proportion given in the table is the cumulation of the ordinates (Y) between Z and the mean: In the normal distribution, the polygon can be considered to consist of extremely narrow categories. That is, the mathematical model describes a hypothetical variable that is *continuous*. In a continuous variable there is an infinite number of categories. Only some of these are ordinarily included in the construction of a histogram. Thus, the normal distribution can be thought of as a polygon constructed from a histogram that consists of categories with zero width. Wherever a line is drawn perpendicular to the base, intercepting the upper boundary of the polygon, it corresponds to a category, and therefore to a value Y (*ordinate*) having a fixed frequency.

There is, theoretically, an infinite number of possible Z values. However, you will note that 99.98 per cent of the area is included between ±4.00Z.

It is, therefore, not necessary to list values of Z greater than those which include practically all the area.

At the center of the distribution, i.e. at the mean μ, Z is 0.00. The proportion of the area to the right of zero Z is .50 and to the left it is also .50. As the distance of a hypothetical X value, in Z units, increases from the mean, the proportion of the area in the distribution increases: for instance, 2.27 per cent of the area lies to the left of $-2.00Z$, and 97.63 per cent of the area lies to the left of $+2.00Z$. Proportions and per cent can be converted to *cases* (X), or a specific number of observations, from the area.

If there are 100 observations in a distribution that is normal in form, the area of the polygon represents 100 per cent of the observations. Then, obviously, 10 per cent of the area represents 10 per cent of the cases. If a variable is normally distributed, you would expect 15.86 per cent of the observations to fall to the left of any value, or category, $-1.00Z$ from the mean. Because all the observations in a distribution can be converted to z, and because under the specified conditions z is approximately equal to Z, proportions of the distribution can be readily determined.

Example: A student obtained a score of 120 on a test. The mean and the standard deviation of the population to whom the test was administered are 100 and 10, respectively. What can be said about the student's performance?

The obtained score (X) can be converted to a Z score, i.e., a standard score:

$$Z = \frac{X - \mu}{\sigma}$$

$$= \frac{120 - 100}{10}$$

$$= +2.00$$

The student's performance is two standard deviations above the mean performance. His Z score is $+2.00$. Referring to Table A.2, we find that 97.63 per cent of the area in the distribution lies to the left of $+2.00Z$, and we can state that 97.63 per cent of the persons who have taken this test have not done as well as this particular student.

If all possible values in a distribution were transformed to Z, like z, then the new distribution, as previously noted, would be a distribution of Z, with a mean $\mu = 0$ and a standard deviation $\sigma = 1.00$. Thus a Z score of 1.00 is one standard deviation above the mean and a Z score of -2.00 falls two standard deviations below the mean of the distribution.

Because proportions of the area, or of the population, between Z values is fixed irrespective of what variable is transformed to Z, so long as the variable is very nearly normally distributed, the transformed scores, Z, are often called standard scores.

Standard Scores

The relationship between standard scores is also shown in Figure 6.4. There are other standard scores besides Z. The statistic T, for instance, is similar to Z. T is a *standardized normal score*. The transformation of the observations in a distribution to T results in a T distribution with a mean of 50, and a standard deviation of 10.

The simplest transformation to T is

$$T = 50 + Z(10)$$

For instance, a student obtains a test score of 85. The mean and standard deviation of the test are 100 and 15, respectively. His Z score is

$$Z = \frac{85 - 100}{15}$$

$$= -1.00$$

and his T score is

$$T = 50 + (-1.00)(10)$$

$$= 50 + (-10)$$

$$= +40$$

The Sampling Distribution of $\bar{X}$

It has been pointed out that *statistics* summarize samples and *parameters* summarize populations. In general practice, samples are obtained in order to make certain inferences about the population from which they came. For instance, it might be desirable to know the value of the mean, μ, of a given population. This can be done by obtaining a representative sample of n observations from the population and noting the mean of that sample. It may then be inferred that the mean of the population, μ, must be similar to the empirically obtained mean of the sample, $\bar{X}$.

In order to make inferences about population parameters from the sample statistics, it is necessary to determine the extent to which one may rely on the information provided by samples.

If a sample is free from bias—that is, if the observations were obtained independently, at random, from the population—the sampling error should be relatively small. In that case, the sample statistic should be a reasonable estimate of the population parameter.

In a sample, the observations are differentially distributed among the scale categories. This distribution is the *distribution of the sample*. The distribution of the sample represents the distribution of the individual observations in the sample of size n.

Occasionally, we may wish to obtain several samples. Let us say we obtain k samples of size n in order to replicate our observations of the sample statistic $\bar{X}$. It is possible to examine this distribution of statistics in the same way that we examined the distribution of a set of observations.

A distribution of statistics is called a *sampling distribution*. The distribution of the statistic $\bar{X}$ is the sampling distribution of $\bar{X}$. The student should familiarize himself thoroughly with the difference between the distribution of a sample and a sampling distribution. This is often a source of confusion.

The best estimate of the population mean (μ) that can be empirically obtained is the mean of the sampling distribution of $\bar{X}$, or the *mean of the means* ($\bar{X}_{\bar{X}}$). Given k number of samples, and k number of sample means, the mean of this distribution can be determined in the same way as the mean of a sample:

$$\bar{X}_{\bar{X}} = \frac{\Sigma \bar{X}}{k} \simeq \mu$$

In the illustrative example below the data consist of 25 sample means from Table 5.1(b), p. 65. The *variance* ($S_{\bar{X}}^2$) and *standard deviation* ($S_{\bar{X}}$) of the *sampling distribution* of $\bar{X}$ are given in Table 6.1.

The standard deviation of the sampling distribution of $\bar{X}$, i.e. $S_{\bar{X}}$, is called the *standard error of the means*. $S_{\bar{X}}$ has the same relationship to the sampling distribution as S has to the distribution of a sample.

When the number of samples (k) and the sample size (n) in a sampling distribution are sufficiently large, the frequency polygon representing that *sampling distribution* will be approximately normal in form (even if the samples were obtained from a distribution that is not normal).

The sampling distribution of $\bar{X}$ has, then, the same properties as any normally distributed population, and it can be used to obtain proportions. The deviations of a sample mean, $\bar{X}$, from the mean of the distribution, $\bar{X}_{\bar{X}}$, i.e. $\bar{X} - \bar{X}_{\bar{X}}$, can be converted to Z by dividing it by the standard error.

When k samples of size n are obtained, the distribution of the k sample means has a mean $\bar{X}_{\bar{X}} = \mu$ and a standard deviation $S_{\bar{X}} = \sigma_{\bar{X}}$, where

$$\sigma_{\bar{X}} = \frac{\sigma}{\sqrt{n}}$$

provided that the samples are reasonably large (greater than 30).

$\bar{X}$	$\bar{X}^2$	
24.4	595.36	$\bar{X}_{\bar{X}} = \dfrac{\Sigma \bar{X}}{k}$
20.6	424.36	
22.3	497.29	
23.2	538.24	$= \dfrac{592}{25}$
24.6	605.16	
23.9	571.21	
20.6	424.36	$= 23.68$
22.8	519.84	
24.6	605.16	$\Sigma x^2 = \Sigma \bar{X}^2 - \dfrac{(\Sigma \bar{X})^2}{k}$
23.0	529.00	
24.0	576.00	
22.8	519.84	$= 14{,}080.64 - \dfrac{(592)(592)}{25}$
26.8	718.24	
24.2	585.64	
24.4	595.36	$= 14{,}080.64 - 14{,}018.56$
23.7	561.69	
21.0	441.00	$= 62.08$
24.1	580.81	
24.8	615.04	$S_{\bar{X}}{}^2 = \dfrac{62.08}{25}$
23.8	566.44	
24.4	595.36	
25.9	670.81	$= 2.48$
21.5	462.25	
24.3	590.49	$S_{\bar{X}} = 1.57$
26.3	691.69	
$\Sigma = 592.0$	$\Sigma = 14080.64$	

TABLE 6.1

When the standard deviation of the distribution is not known, $S_{\bar{X}}$ can be used as an estimate of $\sigma_{\bar{X}}$:

$$S_{\bar{X}} = \frac{S}{\sqrt{n}}$$

Since the sampling distribution of $\bar{X}$ can, ordinarily, be assumed to be normal, conversion of the relevant values to Z provides a distribution whose proportions are fixed and known. The typical transformation is of the form

$$Z = \frac{\bar{X} - \mu}{\sigma_{\bar{X}}}$$

In Table 6.1, the computed $S_{\bar{X}}$ of the sampling distribution, with $k = 25$ sample means, $n = 10$ sample size, is 1.57. The standard deviation of the population from which the samples were obtained is 5.03, and

$$\sigma_{\bar{X}} = \frac{\sigma}{\sqrt{n}}$$

$$= \frac{5.03}{\sqrt{10}}$$

$$= \frac{5.03}{3.16}$$

$$= 1.59$$

The difference between the standard error computed from a sampling distribution of 25 means, of samples of 10 observations (1.57) and the standard error of such a sampling distribution estimated from the parameter, σ, is less than 5 per cent. Thus it may be concluded that $S_{\bar{X}}$ is a reasonable estimate of $\sigma_{\bar{X}}$.

In Figure 6.4, 68.26 per cent of the observations fall between $+1.00$ and $-1.00Z$ in the normal distribution, 95.44 per cent fall between $\pm 2.00Z$, and 99.72 per cent between $\pm 3.00Z$. Since the normal distribution also applies to the sampling distribution of means, you would expect that you would find the same proportion of sample means between corresponding Z units.

Assuming for the moment that the population from which the data in Table 6.1 were obtained is, in fact, normally distributed, we can reconstruct the sampling distribution, given the following information: $\bar{X} = 23.68$, $S_{\bar{X}} = 1.57$. This distribution is shown in Figure 6.5.

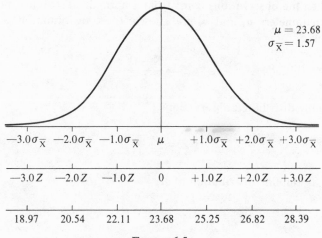

FIGURE 6.5

Given the parameters in Figure 6.5, it can be determined that 95 per cent of sample means have values ranging between 26.82 and 20.54, and 99 per cent of sample means have values ranging between 28.39 and 18.97. The sampling distribution of the means is a valuable tool for testing hypotheses about populations that are based on sample statistics, as we shall see later.

Frequently, the population standard deviation is not known for a given distribution. Hypotheses about the population may be based on statistics summarizing relatively small samples: n = 30 or less.

The sampling distribution of $\bar{X}$ becomes markedly platykurtic as the size of the samples decreases. Because this departure is of some significance when samples are very small (n = 30 or less), proportions of the sampling distribution of $\bar{X}$ based on small samples are given in Table A.4, pp. 267.

When proportions of the sampling distribution of $\bar{X}$ are needed and the standard error, $S_{\bar{X}}$, is based on samples of less than 30 observations, the small sample statistic t is used instead of the ordinary Z.

The Small Sample Statistic t

The smaller the sample size, n, in the sampling distribution, the greater the departure from normal. This departure is toward a platykurtic distribution, and, consequently, toward greater data variability. Ignoring for the moment, the precise shape of the small samples distribution, let us examine what happens to the variability of some samples, obtained at random from a normally distributed population, in a sampling distribution.

Previously, we defined a transformation of individual observations to deviations from the mean, divided by the distribution standard deviation, as z. When the observations come from a normally distributed population, and the parameters μ, and σ, are known, the transformation becomes Z. Thus, where

$$z = \frac{X - \bar{X}}{S}$$

for normally distributed observations,

$$Z = \frac{X - \mu}{\sigma}$$

and, for a sampling distribution of $\bar{X}$,

$$Z = \frac{\bar{X} - \mu}{\sigma_{\bar{X}}}$$

But when the transformation involves a sampling distribution of $\bar{X}$ and df < 30, or when σ is not known and is estimated from the statistic S calculated from a sample with less than 30df,

$$t = \frac{\bar{X} - \mu}{S_{\bar{X}}}$$

The statistic t is a special case of z also. It is a transformation using the sample statistic S to estimate the population parameter σ, from which the standard error of a hypothetical distribution of means, for an infinite number of samples of size n, can be determined. The proportion of the distribution between t units is determined from tables (the t tables) that take into account the differences in sampling distributions based on samples that are relatively small in size.

Figure 6.6 shows the difference between a normal distribution and sampling distributions of 20, 10, 5, and 3 degrees of freedom with regard to the

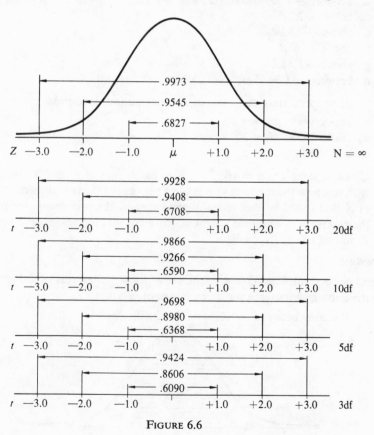

FIGURE 6.6

92   INTRODUCTION TO STATISTICS

proportion of the area between corresponding Z and t units. It should be noted that in the normal distribution, for instance, 99.73 per cent of the area lies between $\pm 3.00Z$, while in the t distribution, with 3df, only 94.24 per cent of the area lies between $\pm 3.00t$. This is a difference of approximately 5.5 per cent. This means that in order to get 99.73 per cent of the area of the distribution with 3df, one would have to go further into the tails of the distribution. Specifically, one would have to go beyond $\pm 8.00t$. In the t distribution with 20df, one would have to go to $\pm 3.4t$ to include the same 99.73 per cent included between $\pm 3.00Z$.

Problems

1. What proportion of the area, in a *normal* distribution, is found:
   a. between -1.2 and $+1.2Z$.
   b. between -2.4 and $+1.6Z$.
   c. between $-1.2Z$ and $0.00Z$, and $+1.2Z$ and $+1.9Z$.
   d. below $-2.5Z$.
   e. above $+1.42Z$.
   f. above $-.63Z$.
   g. above $+1.65Z$.
   h. between $-1.74\,Z$ and $+2.25Z$.

2. What proportion of the area, in a t distribution, is found:
   a. above $1.00t$, df $= \infty$.
   b. above $1.00t$, df $= 20$.
   c. above $1.00t$, df $= 5$.

3. Given a sample: $n = 16$, $\bar{X} = \mu = 25$, $S = 5$, σ is not known. You wish to know the range of values of 95 per cent of the sampling distribution of $\bar{X}$ that could be theoretically generated by the population from which the sample was obtained. How much error would you incur if you used Z instead of t?

Answers

Until one becomes familiar with the use of the Z tables, it is useful to draw a distribution and indicate the areas in question.

1. a. The area between $-1.2Z$ and $+1.2Z$ is .7699.

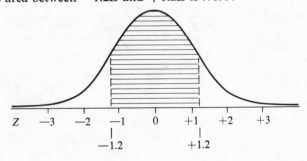

b. The area between $-2.4Z$ and $+1.6Z$ is .9370.

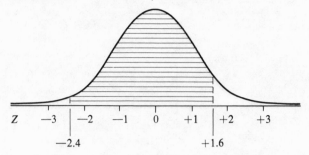

c. The area between $-1.2Z$ and 0.00 and $+1.2Z$ and $+1.9Z$ is .4713. (Note the simpler solution, i.e., $\frac{1}{2}$ the area between $\pm 1.9Z$.)

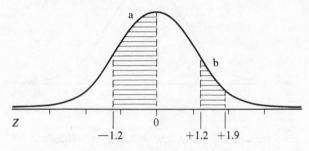

d. The area below $-2.5Z$ is .0062.

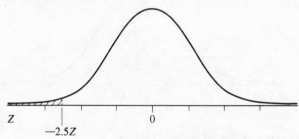

e. The area above $+1.42Z$ is .0779 (by interpolation).

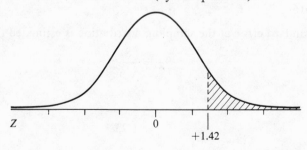

f. The area above $-.63Z$ is .7356.

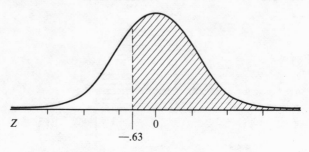

g. The area above $+1.65Z$ is .0495.

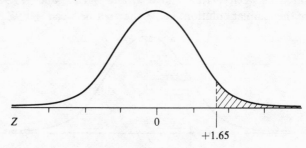

h. The area between $-1.74Z$ and $+2.25Z$ is .9468.

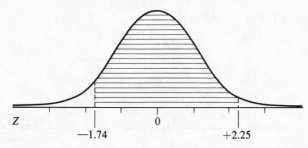

2. a. .1587 (approximated from Z tables).
   b. .1646
   c. .1816

3. The standard error of the sampling distribution is estimated from S; it is

$$S_{\bar{x}} = \frac{S}{\sqrt{n}}$$

$$= \frac{5}{4}$$

$$= 1.25$$

Referring to the $\pm Z$ values in Table A.1, p. 263, which include 95 per cent of the area, we obtain the values $\pm 1.96Z$. And, the t values which include 95 per cent of the area of the sampling distribution based on samples of size 16, is $\pm 2.1t$. With a mean of 25, these values can be used to reconstruct the two distributions:

$$\bar{X} \pm 1.96\,(S_{\bar{X}}) \quad \text{and} \quad \bar{X} \pm 2.1\,(S_{\bar{X}})$$

The actual limits are

$$25 + 1.96\,(1.25) = 22.55$$
$$25 - 1.96\,(1.25) = 27.45$$

and

$$25 + 2.1\,(1.25) = 22.38$$
$$25 - 2.1\,(1.25) = 27.62$$

The error is the difference between the two ranges, since they reflect values for estimated means which would not be included in the estimated 95 per cent of the distribution; this error is 6.5 per cent.

VII

Testing Hypotheses: I

The usefulness of statistics to the investigator depends upon its potential as a research tool to guide him in making meaningful decisions. The decisions he is usually required to make concern the probability of an event, and depend upon the frequency with which that event has been observed previously. The event might be the mean performance of a group of rats in a maze-learning task under specified conditions, or a change in behavior in a group of psychotic patients under the influence of a psychoactive drug agent, or any number of other behavior indices. He may wish to determine how frequently the event is likely to be observed again under similar conditions (replication). Having noted the frequency with which it has already occurred, he must determine how the event relates to all events that might possibly be observed if his observations are *chance observations*. This relationship is described by the probability ratio $\dfrac{E}{E\text{'s}}$.

There are, as you have seen, several different ways in which frequency distributions may differ from each other. They may differ with regard to their form, as indicated by the standard deviation, or with respect to concentration in specific scale categories, i.e. mean, or Q, or median. In order to make reliable decisions, the investigator must recreate, at least conceptually, the frequency distribution of observations from which a particular observed event came: he must know E's as well as E.

This conceptual recreation of a model of a hypothetical frequency distribution involves the following line of reasoning: Given $\bar{X}$, the mean of a sample subject-group performance on a given task, and S, the standard deviation of the distribution of observations. Hypothesis: this group is typical of all groups who perform this given task.

96

The test of this hypothesis depends upon an understanding of what is meant by *typical*. If an infinite number of samples were obtained from a population from which the sample came, and the mean of these samples were plotted, the resulting sampling distribution of $\bar{X}$ could be used to determine whether the empirically observed sample is typical. The specific criteria for establishing that something (a *statistic*) is typical in a distribution will be dealt with in detail later in this chapter. Our concern at the moment is the model for hypothesis testing.

If the mean of the empirical sample is identical to the mean of the hypothetical sampling distribution, we would expect that our sample is typical in the distribution. That is, samples of this kind are common in the hypothetical distribution that defines what we might expect in the long run.

When an event has been observed ($\bar{X}$, for instance) it is possible to establish its probable frequency in a sampling distribution of such events. The investigator generally tests hypotheses concerning the probable frequency of such observed events in terms of the following questions:

1. What is the probability of a specific event E, in a distribution of events, E's?
2. How can the limits indicating all events that are probable in a distribution of such events be determined? That is, what is the value of E_1, E_2, $E_3 \ldots E_n$, all of which are compatible with a given hypothesis?

In the first case we attempt to determine the probability of an event by the $\dfrac{E}{E's}$ relationship and the nature of the distribution of the E's. In the second case we are not so much concerned with a specific E, but rather with a continuum of E hypotheses. We are concerned with all likely and all unlikely events in a distribution according to *a priori* criteria for likelihood.

The hypotheses that investigators may test about empirical data concern either the population parameter or the reliability of the samples. The basic format for these two approaches to hypothesis testing is summarized below:

1. $\bar{X} = \mu$: the observed mean, $\bar{X}$, of a sample is identical to the mean of the population, μ, from which the sample was obtained.
2. $\bar{X}_1 = \bar{X}_2$: the mean of sample 1 is the same as the mean of sample 2. There is the additional implication that if the means of two samples are identical (and the variances are known to be the same) it may be assumed that if the two samples are unbiased, the means of the respective populations from which the two samples came must be the same, i.e. $\mu_1 = \mu_2$, and consequently both samples must have come from the same population. This statement summarizes the hypothesis that two samples are unbiased samples from the same population and is

based on the following contention: if $\bar{X}_1 = \bar{X}_2$, and $\mu_1 = \mu_2$, and if, in addition, $\bar{X}_1 = \mu_1$ and $\bar{X}_2 = \mu_2$, then $\bar{X}_1 - \bar{X}_2 - \mu_1 - \mu_2 = 0$.

3. $S^2 = \sigma^2$: the observed variance of a sample is the same as the variance of the population from which the sample was obtained.

There are other hypotheses; the ones listed above are some of the more basic ones. But all hypotheses take into account the fact that the population parameter is either known or estimated from a sample statistic when the test is based on known proportions of the Z or t distributions.

There are sampling distributions other than the $\bar{X}$ distributions, and they provide the basis for other kinds of tests of hypotheses:

1. $\dfrac{S_1{}^2}{S_2{}^2}$: two samples whose variance ratio is not significantly different from 1.00 come from the same population.

2. There are relationships between population proportions in polychotomous variables and sample proportions of these variables. The difference between empirical proportions and hypothesized proportions is distributed as χ^2 (read chi square).

The first is based upon the F distribution and applies to independently obtained variance estimates from samples that differ in size and number. The second is based on the chi square (χ^2) distribution of differences between empirical and hypothetical proportions of distributions whose categories are discrete-variable categories and whose form is not known, or is known not to be a normal distribution.

The Parametric Assumptions

It is important to note that, with the exception of χ^2 and to some extent F, the validity of tests of hypotheses depends on the satisfaction of the following parametric assumptions:

1. the observations were obtained at random . . .
2. independently (they are not correlated) . . .
3. by sampling with replacement . . .
4. from a normally distributed population of observations . . .
5. with mean μ and standard deviation σ.

When these conditions have been met, a statement of the probability that a hypothesis is correct can be made. For this reason, tests involving Z or t are known as *parametric* tests.

Since a statistical proof rests on a statement of probability concerning what will happen *in the long run*, it is quite possible that an investigator will make errors in specific instances. It is understood, of course, that he will be correct more often than not in his decisions. However, on specific instances, a hypothesis *not rejected* as incorrect may in fact be incorrect. Consequently, the criterion for *failing to reject* a hypothesis is also a statement of the number of times the decision to *fail to reject* a given hypothesis is an improper decision.

By convention, investigators are willing to be wrong no more than five out of 100 times, i.e. 5 per cent of the time. Some investigators choose 10 per cent, while others choose 1 per cent. But since the most frequently used criterion in the behavior sciences is 5 per cent, we shall use it most frequently in this text. The criterion for *failure to reject* or *rejecting* a hypothesis is called the *confidence level*, or the *level of significance*, and is represented by α (Greek alpha).

The Level of Significance: α

If an event, E, occurs less than 5 per cent of the time in a distribution of events, E's, we *reject* the hypothesis that this is a likely or typical event. The format for stating this proposition is called the *null-hypothesis*. The null-hypothesis is a hypothesis of *no-differences*.

The Null-hypothesis

Let us suppose, for the moment, that we have a sample of size n, with mean $\bar{X}$ and standard deviation S. This sample was obtained from a population with mean μ and standard deviation σ. It is our hypothesis that $\bar{X} = \mu$. And, we know that $S = \sigma$. The implication of the validity of this hypothesis is that the sample is an unbiased representation of the population. The formal analysis of this hypothesis, that $\bar{X} = \mu$, requires the formulation of a null-hypothesis—that is, that there is *no difference* between $\bar{X}$ and μ. The null-hypothesis, H_0, states that

$$\bar{X} - \mu = 0$$

Thus, the empirical hypothesis $\bar{X} = \mu$ has been rewritten as a null-hypothesis.

On the basis of random sampling, we expect that the mean of a sample will seldom be exactly equal to the mean of the population from which the sample came. The sampling distribution of $\bar{X}$ tells us that. If the mean of a sample were always equal to the mean of the population from which it came, the sampling distribution of $\bar{X}$ would not be a normal distribution. Since we know, *a priori*, that $\bar{X}$ will seldom be equal to μ, and therefore $\bar{X} - \mu = 0$ will be a relatively rare occurrence, we really must ask a different question—namely:

If $\bar{X} - \mu$ does not equal zero, what values of $\bar{X} - \mu$ will we be willing to accept as being very nearly zero?

The answer to that question is:

We will accept as being very nearly zero all those values of $\bar{X} - \mu$ that do not *differ significantly* from zero.

Admittedly, this is not a very satisfactory answer. But we can then stipulate that when we say "values . . . that do not *differ significantly* from zero" we mean all those values that, according to the criterion α, have a certain probability.

If the criterion α is 5 per cent, we say that we shall consider all values of $\bar{X} - \mu$ that make up 95 per cent of the distribution as not *significantly different* from zero. On the other hand, those values that occur 5 per cent of the time or less in a distribution are considered to *differ significantly* from zero.

This statement is illustrated in Figure 7.1.

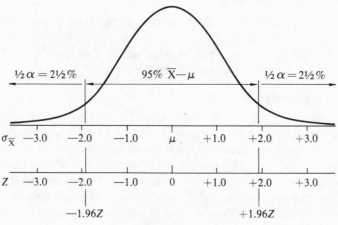

FIGURE 7.1

The Appendix table indicating proportions of the Z-distribution, Table A.1 (p. 263), tells us that the middle 95 per cent of the distribution falls between $\pm 1.96Z$. The other 5 per cent of the distribution is divided equally between the two tails, with 2.5 per cent α in each tail. Thus, the values of $\bar{X} - \mu$ considered not to *differ significantly* from zero are the middle 95 per cent of these values when the criterion is 5 per cent. And all of these middle 95 per cent values are considered to satisfy the contention that

$$\bar{X} - \mu = 0$$

There is an analogy between the probability of selecting a given card from a standard deck of playing cards and the probability of a given value of Z.

In the first case, the probability ratio $\dfrac{E}{E's}$ is $\dfrac{1}{52}$. The event is a given card, and there are 52 possibilities to the outcome of sampling. The probability of Z consists of determining the probability of an event $\bar{X} - \mu$ (E) out of all possible events (E's) defined by the distribution of these events, σ. The difference in complexity between $\dfrac{E}{E's} = \dfrac{1}{52}$ and $\dfrac{\bar{X} - \mu}{\sigma}$ depends upon the nature of the distribution of E's from which the observation in question comes.

The null-hypothesis is *rejected* whenever an event is unlikely, i.e. if it is likely to occur less than α per cent of the time, in a distribution. The rejection of the hypothesis is based on the contention that if a given value of Z can be expected less than α per cent of the time, the difference between $\bar{X}$ and μ, on the basis of which that Z value was obtained, is *significantly different* from zero. If we reject, as unlikely, any event (Z) that occurs less than α per cent of the time, we will be wrong less than α per cent of the time in our decision about a given event. But we will be wrong on some occasion just the same.

In rejecting, or failing to reject, a null-hypothesis, we are always likely to incur an error. This error may be due to the fact that random sampling need not result in identical samples, and in fact seldom does. The probability, in the long run, of being wrong through rejecting (as false) a true hypothesis, is known as an α error (alpha error) because the probability of this error is equal to the value of α.

Type I Errors: α Errors

The probability of rejecting a true hypothesis is equal to the *level of significance*, α. For this reason, errors resulting from rejection of a true hypothesis are known as α errors. The rejection of a true hypothesis occurs α per cent of the time null-hypotheses are rejected.

The α errors are the result of the fact that, when a given $\bar{X} - \mu$ hypothesis is rejected, we state that that value of $\bar{X}$ probably did not come from the particular distribution whose mean is μ because there are only α per cent values smaller and greater. Yet, since there are α per cent values which *are* smaller or greater than the observed event $\bar{X} - \mu$, it *could* have come from that distribution. Consequently, α per cent of the time rejected hypotheses could be true hypotheses.

In the previous hypothetical cases, we have been concerned about the middle portion of the distribution, which does not exceed the limits of α. We might, however, also be concerned with only those α values that exceed a given value. These two approaches are the *confidence limits* (two-tail) approach and the *point estimate* (one-tail) approach to hypothesis testing. These two approaches are outlined in Figure 7.2.

The *confidence limits* approach, or two-tail approach, employs lower and upper limits of a middle (symmetrical about the mean) portion of the distribution. The one-tail approach is a *cumulative area* approach: what proportion of the area lies below a given $\bar{X} - \mu$ value, or what proportion exceeds it? In the two-tail approach, the rejection region for unacceptable hypotheses is symmetrical about the mean, while in the one-tail test, the rejection region is above $1.00 - \alpha$.

Two-tail approach: confidence limits of 95% of the area in a sampling distribution.

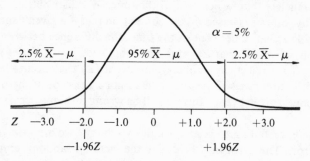

One-tail approach: cumulative area below and above a given point in the distribution.

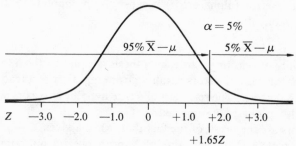

FIGURE 7.2

Alpha errors result from the rejection of null-hypotheses, and their magnitude, in the long run, is α per cent. They occur when events that are statistically unlikely, but that occur nevertheless a specified per cent of the time, are rejected. If you were asked the likelihood of obtaining 50 heads in a 50-coin toss, you would state that this event is improbable but possible. In the long run you would be incorrect in stipulating that 50 heads is a frequent event. However, though this event may be unlikely, it can be expected to occur some per cent of the time the coins are tossed. If you said that you are unwilling to accept the event as probable, you would also be wrong—α per cent of the time.

It is important in this context to make a distinction: the *failure to reject* a null-hypothesis is not the same thing as *accepting* a null-hypothesis. There is no basis for accepting a null-hypothesis. There is no statistical procedure that can be used to predict precisely what will happen on a given occasion. Figure 7.3 shows the frequency distribution of two populations that overlap. In the area of overlap, there is no way of determining whether an observation comes from distribution 1 or distribution 2. Since we cannot differentiate between distributions in this area of overlap, we can at best fail to reject the hypothesis that an observation in distribution 1 actually came from distribution 1.

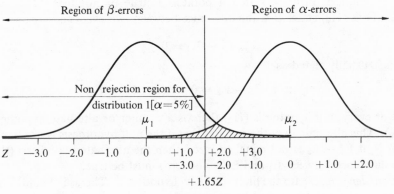

FIGURE 7.3

Type II Errors: β Errors

Every test of a statistical hypothesis contains the possibility of two errors: α and β. While the α errors result from the rejection of true hypotheses, the β errors (beta errors) result from the *failure to reject a false hypothesis.*

These errors result from the fact that, when looking at two distributions that overlap, it is impossible to specify which distribution a particular observation came from. It can be stated with some degree of confidence that an observation came from one particular distribution, but then again, it could have come from the other. Note, in Figure 7.3, that the observations in the non-rejection region of distribution 1 are, more likely than not, actually from that distribution. But, there are also distributions in the non-rejection region of distribution 1 which actually *could* have come from distribution 2.

Clearly, however, an observation which fits into a distribution 1 category would ordinarily not be rejected as unlikely to have come from distribution 1.

If it in fact came from distribution 2, failure to reject that value results in a β error.

There are several ways in which type I and type II errors may be minimized. Alpha errors may be minimized by increasing the criterion limits from, let us say, 5 per cent to 1 per cent. However, in so doing, you increase β errors. Alpha errors may also be minimized by increasing n, the sample size. However, β errors are not automatically decreased when α errors are decreased.

Confidence Limits

There are many types of hypotheses about populations that can be tested, and the criteria for likely and unlikely hypotheses should be clarified further. The use of the null-hypothesis serves to reduce the decision-making process to its simplest form: a stated hypothesis is true or not true. The null-hypothesis (H_0) states that the difference between two events is zero:

$$E_1 - E_2 = 0$$

This is usually stated as

$$H_0: \qquad E_1 - E_2 = 0 \qquad (\text{or} \quad E_1 = E_2)$$

For every null-hypothesis (H_0) there is a counter or alternate hypothesis (H_a). This alternate hypothesis states that if the null-hypothesis is not true—that is, if $E_1 - E_2$ is not equal to zero—then the alternate hypothesis (that the difference between E_1 and E_2 is not zero) must be true.

Confidence limits are the limits of likely hypotheses. The use of confidence limits can best be illustrated by an example.

Let us assume that we have obtained a sample of size 20. The mean is 26.45, and the standard deviation is 10.54. That is all the information we have. We might wish to know the most likely values of μ, the mean of the population from which the sample came. In order to determine what the value of the mean might be, we set up confidence limits.

We begin by assuming that the sample mean is probably an unbiased estimate of the population mean: $\bar{X} \simeq \mu$. The level of significance is arbitrarily 5 per cent, i.e. we wish to be wrong no more than 5 per cent of the time.

A hypothetical sampling distribution of $\bar{X}$ can be recreated to represent the sampling distribution that could be generated from the population from which the sample (n = 20) was obtained. This is done by obtaining an estimate of the standard error of the hypothetical sampling distribution. This estimate is based on the sample statistic S:

$$S_{\bar{X}} = \frac{S}{\sqrt{n}} = \frac{10.54}{4.47} = 2.36$$

It is known that 95 per cent of the area around the mean of a sampling distribution of $\bar{X}$ (which is a normal distribution) is contained between $\pm 1.96Z$. However, since σ is not known, and n is less than 30, we cannot use Z. Therefore we use t. Referring to the t tables, we find that 95 per cent of the area of the sampling distribution with 19df is contained between $\pm 2.09t$.

The lower limit of the confidence limits for μ is

$$\bar{X} - t_{.975}(S_{\bar{X}}) = 26.45 - 2.09(2.36) = 21.52$$

The upper limit of the confidence limits for μ is

$$\bar{X} + t_{.975}{}^{*}(S_{\bar{X}}) = 26.45 + 2.09(2.36) = 31.38$$

Thus we state that 95 per cent of the time we would expect μ to have a value between 21.52 and 31.38.

It is not infrequent, in experimental procedures, to find that we have obtained samples but know little about the parametric values of the population from which the samples come. The confidence limits delineate those values in the distribution that are most likely to be μ. (There are confidence limits for parameters other than μ.)

The confidence limits of μ are based on estimates of the population parameters μ, σ, and the standard error of a hypothetical sampling distribution $\sigma_{\bar{X}}$ for samples of size n. These estimates are the statistics $\bar{X}$, S, and $S_{\bar{X}}$. The precision with which the estimates can be made depends, in part, on the satisfaction of the parametric assumptions.

The confidence limits are the limits of the confidence interval: the region of non-rejection of the null-hypothesis. Say we take any hypothesis of the form $\bar{X} - \mu = 0$ and convert it to t $\left(\text{that is, } t = \dfrac{\bar{X} - \mu}{S_{\bar{X}}} \right)$. If the value of t does not exceed the limits of the confidence interval, then the hypothesis, $\bar{X} - \mu = 0$, will not be rejected.

Point Estimates

It is not necessary to test all possible values that might be the parameter μ. A *point estimate* of μ can be used. A point estimate of a parameter is a specific hypothesis about the value of the parameter.

For instance, given the information in the example above. Could the mean of the population from which that sample came be 33.5?

---

* $t_{.975}$ rather than $t_{.95}$: the t tables are one-tail tables.

This hypothesis is tested in the following way:

$$t = \frac{\bar{X} - \mu}{S_{\bar{X}}}$$

$$= \frac{26.45 - 33.50}{2.36}$$

$$= \frac{7.05}{2.36}$$

$$= -2.99$$

The minus sign can be ignored since the distribution is symmetrical. Referring to the t tables for 19df in the Appendix, we find that the value 2.093 corresponds to the limits for 95 per cent of the area of the sampling distribution. This value is a *critical ratio:* any value of t that exceeds it is outside the confidence limits and is thus *rejected*. The obtained value of t, 2.99, is in the rejection region and we conclude that it is unlikely that, if the sample meets the parametric assumptions, the value 33.50 could be μ, the mean of the population from which that sample was obtained.

Problems

1. Given the following 50 observations:

$$
\begin{array}{ccccc}
17 & 37 & 9 & 35 & 11 \\
11 & 8 & 22 & 20 & 27 \\
18 & 25 & 24 & 29 & 14 \\
27 & 7 & 15 & 10 & 19 \\
6 & 47 & 16 & 22 & 32 \\
22 & 23 & 31 & 21 & 15 \\
34 & 1 & 26 & 11 & 33 \\
13 & 21 & 14 & 35 & 28 \\
21 & 25 & 16 & 1 & 26 \\
29 & 23 & 19 & 16 & 30 \\
\end{array}
$$

   a. Could the mean of the population from which the sample came be 30.00?
   b. Could it be 16.50?
   c. Could it be 36.27?
   d. Could it be 21.00?
   e. What are the 95 per cent confidence limits of μ?
   f. What are the 99 per cent confidence limits of μ?

Answers

1. a. It must first be determined that $\bar{X} = 20.84$,

$$S = 9.58, \text{ and } S_{\bar{X}} = 1.36.$$

$$H_0: \quad \bar{X} - \mu = 0, \quad \alpha = .05 \quad \text{(arbitrarily)}.$$

$$t = \frac{\bar{X} - \mu}{S_{\bar{X}}}$$

$$= \frac{20.84 - 30.00}{1.36}$$

$$= -6.74$$

$$P(t = -6.74) < 5 \text{ per cent, at 49df.}$$

Thus, it is not likely that 30.00 could be the mean of the population from which the sample came.

b. $t = 3.19$, reject the null-hypothesis.

c. $t = -11.35$, reject the null-hypothesis.

d. $t = -0.12$, fail to reject the null-hypothesis.

e. 95 per cent confidence limits of μ are:

$$\bar{X} - t_{\frac{1}{2}\alpha}(S_{\bar{X}}) < \mu > \bar{X} + t_{1-\frac{1}{2}\alpha}(S_{\bar{X}}) = 20.84 - 2.01(1.36) = 18.11$$

and

$$20.84 + 2.01(1.36) = 23.57$$

f. 99 per cent confidence limits of μ are:

$$20.84 - 2.68(1.36) = 17.20$$

and

$$20.84 + 2.68(1.36) = 24.48$$

VIII

Testing Hypotheses: II

In the previous chapter we dealt with general aspects of statistical inference. In this chapter we will discuss statistical procedures that are specific to the hypotheses one may test about populations and samples. The main purpose of these hypotheses is to determine something about the population from which the samples were obtained.

We have discussed the level of significance, α, the probability of the parameter Z, and the statistic, t. It might be well to digress for a moment, and to examine again the conceptual relationship between simple probability ratios, $\dfrac{E}{E\text{'s}}$, encountered earlier, and the more complex ratios Z and t.

In the simple probability ratio, the probability of the event E can be determined by establishing the ratio of the frequency of events in a particular category to the sum of the events in all categories. In the simple probability ratio, the frequency of the events in each category is 1.00; thus the sum of the frequencies is the total number of categories in which there is at least one observation. In more complex probability ratios, such as Z and t, the ratio provides a numerical value which does not, in itself, represent probability, but is a value for which the probability is given in a table. All Z and t values comprise a derived, or indirect, scale of the distance of an event, E, from the mean of the distribution, μ. In the case of Z, for instance, the distance $\dfrac{X - \mu}{\sigma}$ can be used to determine the height of the ordinate corresponding to the category in which X is contained. Each ordinate, Y, has a fixed frequency in the normal distribution. We are told that, for a particular X observation, the category in which that X is observed has an ordinate indicating that the frequency of that category is 20. The distribution, let us say,

108

contains 100 observations. We are asked to determine the probability of obtaining an observation from that category on the basis of random sampling. There are, over-all, 100 observations in the distribution, and 20 observations in a particular category. The probability of selecting an observation from that particular category must be $\frac{20}{100}$, or .20. That is *essentially* how the Z and t tables provide probability estimates.

Hypotheses About the Population Mean: μ

When a sample has been obtained empirically, it is sometimes necessary to determine the nature of the population from which it came. It may be desirable to estimate the value of the population parametric values.

X
11
12
26
28
30
48
51
72
95
120
112
87
90
93
64
33
34
35
30
30

This sample was obtained from a population with
$\mu = 50.00$
$\sigma = 30.00$
$\sigma_{\bar{X}} = 6.71$

$\Sigma X = 1101$
$n = 20$
$\bar{X} = 55.05$
$S^2 = 1145.10$
$S = 33.84$
$S_{\bar{X}} = 7.57$

TABLE 8.1

Samples are seldom obtained without a specific purpose. Frequently they are obtained for just the purpose of determining population parameters, though investigators would prefer to base their information on the entire population. But this is seldom practical, or even possible. In point of fact, with adequate sampling techniques, it is not necessary.

Estimating μ when σ is known

When estimating a population parameter, it is always advisable to use whatever information is available about the population. This procedure greatly enhances the precision of the estimate.

For the estimates here we will use the sample of 20 observations given in Table 8.1. Note that the mean and standard deviation of the population from which this sample came is also given.

Suppose that we did not know the mean of the population, μ, and we wished to estimate it from the sample mean, $\bar{X}$. We may test a specific hypothesis (point estimate), or obtain the confidence limits for μ.

Point Estimate of μ

The point estimate is, essentially, a guess concerning the probable value of μ, with a statistical test to determine whether the guess is reasonable. For instance, we might guess that $\mu = 57.00$. The test consists of verifying the null-hypothesis:

$$H_0: \quad \bar{X} - \mu = 0$$
$$H_a: \quad \bar{X} - \mu \neq 0$$
$$\alpha = .05$$
$$Z = \frac{\bar{X} - \mu}{\sigma_{\bar{X}}}$$
$$= \frac{55.05 - 57.00}{6.71}$$
$$= -.29$$
$$P(Z = -.29) = .3821$$

Since .38 is greater than .05, we fail to reject the null-hypothesis. This does not mean, however, that $\mu = 57.00$. It simply means that it could be that $\mu = 57.00$.

Given the same information, we can also compute the limit of the values that μ might be.

Confidence Limits of μ

The 95 per cent confidence limits of μ are

$$\bar{X} - Z_{\frac{1}{2}\alpha}(\sigma_{\bar{X}}) \quad \text{and} \quad \bar{X} + Z_{1-\frac{1}{2}\alpha}(\sigma_{\bar{X}})$$

Substituting, we get

$$55.05 - 1.96(6.71) = 55.05 - 13.15 = 41.90$$

and

$$55.05 + 1.96(6.71) = 55.05 + 13.15 = 68.20$$

These limits contain 95 per cent of the sampling distribution of $\bar{X}$, for sample of size 20, and are correct if the obtained sample $\bar{X} = 55.05$ is unbiased. From the difference between $\bar{X}$ and μ, you can see that a small sampling error obtains.

Estimating μ when σ Is Not Known

More often than not, when μ is not known, σ is also not known. Some investigators use the procedure above and the probability of Z, even when σ is not known, provided the sample size is greater than 30. Strictly speaking, the procedure is acceptable though not exactly correct. The t statistic should then be used. To illustrate its use, data in Table 8.1 are employed.

Point Estimate of μ

$$\begin{aligned}
H_0: &\quad \bar{X} - \mu = 0 \\
H_a: &\quad \bar{X} - \mu \neq 0 \\
\alpha &= .05 \\
t &= \frac{\bar{X} - \mu}{S_{\bar{X}}} \\
&= \frac{55.05 - 57.00}{7.56} \\
&= -.26
\end{aligned}$$

$P(t = -.26)$ is more than .05. Thus we fail to reject the null-hypothesis.

Confidence Limits of μ

The 95 per cent confidence limits of μ are

$$\bar{X} - t_{\frac{1}{2}\alpha}(S_{\bar{X}}) = 55.05 - 2.09(7.56) = 39.25$$

and

$$\bar{X} + t_{1-\frac{1}{2}\alpha}(S_{\bar{X}}) = 55.05 + 2.09(7.56) = 70.85$$

The value 2.09 is obtained from the t table for sampling distributions of t with 19df. The notation $t_{\frac{1}{2}\alpha}$ (also $Z_{\frac{1}{2}\alpha}$) indicates that half the area of rejection is on the left side (the left-hand tail), and the notation $t_{1-\frac{1}{2}\alpha}$ (also $Z_{1-\frac{1}{2}\alpha}$) indicates that the other half of the rejection region is in the other (upper) tail.

In the context of confidence limits, it should be recalled that "compatible hypotheses" are all hypotheses that are statistically likely, i.e. all the hypotheses which will not be rejected at the α per cent level of significance. All the hypotheses in the confidence interval—between the limits—are hypotheses that meet the criteria for non-rejection.

Thus, the level of significance identifies that portion of the distribution which consists of improbable events. But the rationale underlying the test of hypotheses, the parametric assumptions, must be satisfied. When an empirical sample has been obtained, and a parameter is estimated, a test of the probability that the hypothesized parameter is of the correct value depends upon the assumption that the parametric assumptions *have* been satisfied: rejection of the null-hypothesis does not always reflect *real* differences between statistic and parameter. In fact, rejection of the null-hypothesis may also indicate that one of the parametric assumptions has not been met; the sample may not have come from a normally distributed population, etc.

It should be emphasized that when we are reasonably certain that we have unbiased samples, the statistics $\bar{X}$ and S should be reasonably good estimates of μ and σ. If this were not the case, there would hardly be any point in sampling and testing hypotheses.

Hypotheses About the Population Variance: σ^2

Estimating the mean of a population depends on the sampling distribution of $\bar{X}$. Likewise, estimating the variance of a population is also based on a sampling distribution: the sampling distribution of S^2. The two sampling distributions are quite different in nature.

An estimate of the probable value of μ is based on percentiles of the sampling distribution of $\bar{X}$, as indicated by $Z = \dfrac{\bar{X} - \mu}{\sigma_{\bar{X}}}$, and an estimate of the probable value of σ^2 is based on the sampling distribution of $\dfrac{S^2}{\sigma^2}$. The distribution of $\dfrac{S^2}{\sigma^2}$ is a $\dfrac{\chi^2}{df}$ distribution (read chi square over degrees of freedom). Proportions of this distribution are given in Table A.5(b), p. 269.

Confidence Limits of σ^2

The $1 - \alpha$ per cent confidence limits for σ^2 are

$$\frac{S^2}{P_{\frac{1}{2}\alpha}} \quad \text{and} \quad \frac{S^2}{P_{1-\frac{1}{2}\alpha}}$$

Given the information in Table 8.1, let us assume that σ^2 is not known, and that we wish to estimate it from the sample statistic S^2.

The 95 per cent confidence limits for σ^2 are

$$\frac{S^2}{.469} \quad \text{and} \quad \frac{S^2}{1.73}$$

where the values in the denominator are the 2.5 per cent and 97.5 per cent values (listed in the $\frac{\chi^2}{df}$ tables) needed to obtain the middle 95 per cent of the sampling distribution. Substituting our data, we obtain

$$\frac{1088}{.469} = 2319.83 \quad \text{and} \quad \frac{1088}{1.73} = 628.90$$

The 95 per cent confidence limits for σ are simply the square roots of the confidence limits for σ^2:

$$\sqrt{2319.83} \quad \text{and} \quad \sqrt{628.90}$$

Hypotheses About the Variance of Two Populations

In the following section, we shall examine the hypotheses that can be tested about the mean of two populations. One of the requirements for these tests is the determination that the variances of the two populations are not significantly different. Let us consider the following example: two samples have been obtained from a normally distributed population. These samples consist of n observations on an experimental group and n observations on a control group. You wish to determine whether the experimental treatment has had an effect. A *treatment effect* may be manifest in differences in mean or in variance (and hence in standard deviation). The reason for this is simple. The mean reaction time of a population of subjects under the influence of alcohol, for instance, is different from the mean reaction time for a population of sober subjects. Thus, a sample of subjects from a sober population would, if given alcohol, belong to a new population: the population of subjects under the influence of alcohol. Samples of each of these populations should reflect the change.

Determing that two samples have different variances indicates that they do not belong to the same populations. Since we are concerned with whether two samples could have come from, or belong to, the same population, it is sufficient that we know they differ significantly in form in order to know that they are not from the same population.

Given two samples, sample 1 has a mean $\bar{X}_1$, a variance $S_1{}^2$, and n_1 observations. Sample 2 has a mean $\bar{X}_2$, a variance $S_2{}^2$, and n_2 observations.

If these two samples are obtained at random from a normally distributed population, the ratio of their variances is distributed as F:

$$F = \frac{S_1{}^2}{S_2{}^2}$$

where $S_1{}^2$ is the larger and $S_2{}^2$ is the smaller of the two variance estimates.

Percentiles of the F distribution are given in Table A.7, pp. 271–74. The percentiles are given for degrees of freedom in the numerator of the ratio, and degrees of freedom in the denominator.

The F distribution table is not one distribution, but a set of distributions. Each one is the sampling distribution of ratios of pairs of variances obtained at random from a normally distributed population. The F tables are one-tail tables, giving proportions—as it were, percentiles—of the area of the distribution, below standard critical ratios. Ordinarily, the F ratio is constituted this way:

$$F = \frac{\text{larger } S^2}{\text{smaller } S^2}$$

The F test is a test for *homogeneity of variances*, that is, similarity of distributions. If the variances of two populations are identical, the ratio between them is 1.00. If the variances of two populations are similar, their ratio should not differ significantly from 1.00. For illustrative purposes, let us examine two samples whose variances are, respectively, 194.91 (9df), and 85.11 (9df). The ratio of their variances are

$$F = \frac{\text{larger } S^2}{\text{smaller } S^2}$$

$$= \frac{194.91}{85.11}$$

$$= 2.29$$

The level of confidence is 5 per cent.

The probability of F = 2.29, df_1, df_2, is greater than 5 per cent and we fail to reject the hypothesis that $S_1{}^2 = S_2{}^2$. In this example, the respective df's are 9 and 9. In Table A.7 the value given at the 5 per cent level of significance is 3.18. This value is the critical ratio: an F ratio that exceeds the table value is significant at the α per cent level.

You will note from Table A.7 that the degrees of freedom need not be identical for the two samples. The F tables tell us what values resulting from the ratio of two variances, from samples of various sizes, are likely to occur on the basis of chance, and what values are not likely to occur on that basis.

Hypotheses About the Mean of Two Populations: μ_1 and μ_2

If we were to pair, at random, a large number of sample means, obtained at random from the same population, differences between these pairs of means will result in a distribution: the distribution of differences whose mean is $\bar{X}_{d\bar{x}}$, and whose standard error is $S_{d\bar{x}}$, the standard error of the differences.

From a population with $\mu = 24.35$, and $\sigma = 10.07$, we have obtained two samples:

Sample 1	Sample 2
$\bar{X}_1 = 25.80$	$\bar{X}_2 = 23.00$
$S_1 = 13.96$	$S_2 = 9.23$
$n_1 = 25$	$n_2 = 25$

These data provide the basis for the illustrative examples in the sections below.

Differences Between Means of Independent Populations when σ Is Known and $\sigma_1 = \sigma_2$

We wish to determine whether the observed differences between the two sample means warrants the assumption that this difference is due to sampling error. If the difference cannot be attributed to sampling error, then it is possible that the two samples came from different populations. If the two samples did, in fact, come from the same population, the difference between their means would be significantly different from zero only α per cent of the time, on the basis of chance alone.

The null-hypothesis expressing the difference between the two sample means is

$$H_0: \quad \bar{X}_1 - \bar{X}_2 = 0$$

and the alternate hypothesis is

$$H_a: \quad \bar{X}_1 - \bar{X}_2 \neq 0$$

The level of significance is 5 per cent.

It should be noted that if

$$\bar{X}_1 - \mu_1 = 0, \quad \text{i.e. } \bar{X}_1 = \mu_1$$
$$\bar{X}_2 - \mu_2 = 0, \quad \text{i.e. } \bar{X}_2 = \mu_2$$
$$\bar{X}_1 - \bar{X}_2 = 0, \quad \text{i.e. } \bar{X}_1 = \bar{X}_2$$

then

$$\bar{X}_1 - \mu_1 - \bar{X}_2 - \mu_2 = 0$$

The latter statement stipulates that we are really testing to see whether the two samples came from the same population.

The test of this hypothesis is

$$Z = \frac{\bar{X}_1 - \bar{X}_2}{\sigma \sqrt{\left(\dfrac{1}{n_1}\right) + \left(\dfrac{1}{n_2}\right)}}$$

The operation in the denominator of the ratio results in the geometric mean of the error terms. The geometric mean of two quantities is the square root of their products: $\sqrt{(\bar{X}_1) + (\bar{X}_2)}$.

Substituting the given data, we obtain

$$Z = \frac{25.8 - 23.0}{10.07 \sqrt{\left(\dfrac{1}{25}\right) + \left(\dfrac{1}{25}\right)}}$$

$$= \frac{25.8 - 23.0}{10.07 \sqrt{.08}}$$

$$= \frac{2.80}{2.85}$$

$$= .98$$

The probability that $Z = .31$ is greater than 5 per cent, and we fail to reject the null-hypothesis.

Differences Between Means of Independent Populations when σ^2 Is Not Known

Under certain circumstances, it is desirable to test the hypothesis that $\bar{X}_1 - \bar{X}_2 = 0$ when the population variance is not known. The level of significance is, arbitrarily, 5 per cent.

$$H_0: \quad \bar{X}_1 - \bar{X}_2 = 0$$
$$H_a: \quad \bar{X}_1 - \bar{X}_2 \neq 0$$
$$\alpha = 5 \text{ per cent}$$

The test is

$$t = \frac{\bar{X}_1 - \bar{X}_2}{S_p \sqrt{\left(\dfrac{1}{n_1}\right) + \left(\dfrac{1}{n_2}\right)}}$$

S_p is the square root of $S_p{}^2$, the *pooled variance*, and is defined as

$$\frac{(n_1 - 1)S_1{}^2 + (n_2 - 1)S_2{}^2}{n_1 + n_2 - 2}$$

This is the *pooled standard deviation*.

Before the variances can be combined to form the S_p term, the variances must be shown to be *homogeneous*. This can be determined by the use of the F test:

$$F = \frac{\text{larger } S^2}{\text{smaller } S^2}$$

substituting our data, we obtain

$$F = \frac{194.88}{85.19}$$

$$= 2.29$$

Using Table A.7 for percentiles of the F distribution for $df_1 = 24$, $df_2 = 24$, we find that $F = 2.29$ is greater than the critical value. The hypothesis that $S_1{}^2 = S_2{}^2$ is rejected, and there is no need to proceed further to reject the hypothesis that the two samples came from the same population.

If we wished to test the differences between the populations to determine whether the differences between the means is significant, the procedure would be

$$t = \frac{\bar{X}_1 - \bar{X}_2}{\sqrt{\left(\dfrac{S_1{}^2}{n_1}\right) + \left(\dfrac{S_2{}^2}{n_2}\right)}}$$

Substituting our data, we obtain

$$t = \frac{25.80 - 23.00}{\sqrt{\left(\dfrac{194.88}{25}\right) + \left(\dfrac{85.19}{25}\right)}}$$

$$= \frac{2.80}{3.35}$$

$$= .84$$

$P(t = .84)$ is greater than 5 per cent, and we fail to reject the null-hypothesis. Here, then, is a case of two populations with similar means but different variances. The t distribution, above, has

$$\frac{\left[\left(\dfrac{S_1{}^2}{n_1}\right) + \left(\dfrac{S_2{}^2}{n_2}\right)\right]^2}{\dfrac{\left(\dfrac{S_1{}^2}{n_1}\right)^2}{n_1 + 1} + \dfrac{\left(\dfrac{S_2{}^2}{n_2}\right)^2}{n_2 + 1}} - 2 \text{ degrees of freedom} = 40df$$

Figure 8.1 illustrates distributions that are similar in form but different in means, and distributions that are different in form but similar in means.

Differences Between Means of Independent Populations when the Observations Are Paired

Investigators sometimes wish to determine the differences between two groups which may be ascribable to some specific treatment. However, the groups are different to begin with. That is, there might be an extraneous

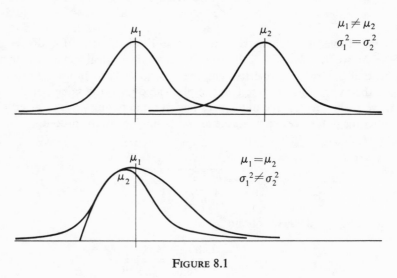

FIGURE 8.1

variable operating irrespective of treatment effect. This extraneous variable may account for the differences between the groups.

The effect of instructional method on school children learning mathematics, for instance, given two different methods, is difficult to assess because often there are differences already existing between school classes before the study is begun. Under these conditions, it might be better to use *matched groups*, as compared, for instance, to random samples.

The effect of different methods of instruction on groups that are different to begin with can be determined by studying pairs of students matched for a variety of relevant factors, including age, general ability as measured by standard tests, etc.

If we denote the first group in such a study as group a and the second group as group b, we can obtain pairs of subjects matched and derive a distribution consisting of observations of the difference between the measures obtained from each member of the pair: that is, the difference between members of the a and b groups. This distribution is outlined below:

$$d_1 = X_{a_1} - X_{b_1}$$
$$d_2 = X_{a_2} - X_{b_2}$$
$$d_3 = X_{a_3} - X_{b_3}$$
$$\cdot \qquad \cdot \qquad \cdot$$
$$\cdot \qquad \cdot \qquad \cdot$$
$$\cdot \qquad \cdot \qquad \cdot$$
$$d_n = X_{a_n} - X_{b_n}$$

To test the hypothesis that $\mu_a = \mu_b$, it is not necessary to assume that $\sigma_a^2 = \sigma_b^2$.

Let us assume that the illustrative example below consists of two groups which it was found to be desirable to pair with respect to some criterion. The distribution of the differences between the matched pairs of observations is given in Table 8.2.

Pair	Group a	Group b	Difference (d)	d^2
d_1	39	16	23	529
d_2	22	21	1	1
d_3	10	22	12	144
d_4	19	16	3	9
d_5	12	19	7	49
d_6	23	14	9	81
d_7	33	4	29	841
d_8	55	29	26	676
d_9	31	31	0	0
d_{10}	14	18	4	16

$$\Sigma X_a = 258 \quad \Sigma X_b = 190 \quad \Sigma d = 114 \quad \Sigma d^2 = 2346$$
$$\bar{X}_a = 25.8 \quad \bar{X}_b = 19.0 \quad \bar{d} = 11.4$$

$$\Sigma x_d^2 = \Sigma d^2 - \frac{(\Sigma d)^2}{n}$$

$$= 2346 - \frac{(114)(114)}{10}$$

$$= 2346 - \frac{12996}{10}$$

$$= 1046.4$$

$$S_d^2 = \frac{\Sigma x_d^2}{n} \qquad\qquad S_d = \sqrt{S_d^2}$$

$$= \frac{1046.4}{10} \qquad\qquad = \sqrt{104.64}$$

$$= 104.64 \qquad\qquad = 10.23$$

TABLE 8.2

The null-hypothesis is

$$H_0: \quad \bar{X}_a - \bar{X}_b = 0$$

and the alternate hypothesis is

$$H_a: \quad \bar{X}_a - \bar{X}_b \neq 0$$

The level of significance, α, is 5 per cent.

The test is

$$t = \frac{\bar{X}_a - \bar{X}_b}{\dfrac{S_d}{\sqrt{n}}}$$
$$= \frac{25.8 - 19.0}{3.24}$$
$$= 2.10$$

The probability of $t = 2.10$, df $= n - 1$ (df $= 9$), is greater than 5 per cent, and thus we fail to reject the null-hypothesis.

General Recommendation for Rejection of the Null-hypothesis

In the tests of significance, Z, t, and F, the 5 per cent level of significance (the confidence level) has been employed in the illustrative examples. This is purely arbitrary. The 1 per cent, or 10 per cent, or, for that matter, the 50 per cent level could have been used. The scientific investigator who is engaged in a study chooses the level of significance. If he is willing to be wrong 50 per cent of the time, that is his decision to make. It is recommended that he choose the level of significance before analyzing his data. That is, he should report the probability of the event studied in terms of whatever the probability of the obtained ratio is, be it greater or smaller than a predetermined criterion for rejection, or failure of rejection, of the null-hypothesis.

It is not infrequent that comprehensive studies in which several comparisons are made present the probability of each comparison in terms of whether it is greater or smaller than 10 per cent, 5 per cent or 1 per cent. That is, comparison 1 may be stated to be significant at the 1 per cent level, while 2 and 3 are significant at the 10 per cent level, and 4 is significant at the 5 per cent level. This is generally not recommended. The criterion for rejection should be stated and the hypotheses should be rejected or not rejected as the occasion demands in terms of one stated α level.

Clearly, all hypotheses are significant at a particular level. Why should an investigator be willing to be wrong only 5 per cent of the time with one hypothesis and 10 per cent of the time with another? This form of report is tantamount to an apology for a failure to find predicted differences between groups.

Note, for instance, that if you obtain a t ratio of 2.21, with 9df, the reflected difference is significant at the 5 per cent level of significance. If the ratio had been 2.1, it would not have attained significance at that level. It would, however, have been significant at the 10 per cent level. But the difference between a t ratio needed to attain significance at the 10 per cent level (1.796) and a t ratio needed to attain significance at the 5 per cent level, with 9df, is only .045 t units. This should make the student reflect on the fact that adapting the level of significance to the obtained probability ratio is a questionable procedure.

It soon becomes manifestly apparent to anyone doing research that results are either significant or not significant. Results that are *almost* significant are, at best, doubtful.

Problems

1. Given the following two samples:

Sample 1	Sample 2
17	23
11	1
18	21
27	25
6	23
22	9
34	22
13	24
21	15
29	16
37	31
8	26
25	14
7	16
47	19

   a. If the mean of the population from which these samples were obtained is 20, and the standard deviation is 10, is the last observation in sample 1 (47) typical of that population? On the basis of chance alone, what is the probability of obtaining such an observation in that population?

   b. If $\sigma = 10$, what is the probability that $\bar{X}_1$ came from a population with $\mu = 20$?

   c. Based on the mean of group 1, with the given population standard deviation, what are the 95 per cent confidence limits of μ?

   d. Based on the sample mean and standard deviation of group 2, what are the 95 per cent confidence limits of μ?

e. With the information in 1b, above, what is the probability that $\mu = 20$?

f. If $\sigma_1 = \sigma_2 = \sigma$, but σ is not known, what is the probability that $\bar{X}_1 = \bar{X}_2$?

g. If $\sigma_1 \neq \sigma_2$, how would you determine the probability that $\bar{X}_1 = \bar{X}_2$?

h. Arrange the observations in each of the two samples in ascending order. Now pair the first, second, third ... nth observation. What is the probability that $\bar{X}_1 - \bar{X}_2 = 0$?

Answers

1. a.
$$Z = \frac{X - \mu}{\sigma}$$
$$= \frac{47 - 20}{10}$$
$$= 2.7$$

$P(Z = 2.7)$ is .0104.

b.
$$Z = \frac{\bar{X} - \mu}{\sigma_{\bar{X}}}$$
$$= \frac{21.47 - 20.00}{10\sqrt{15}}$$
$$= \frac{1.47}{38.70}$$
$$= .04$$

$P(Z = .04)$ is greater than 5%, not an unlikely event.

c. The 95 per cent confidence limits of μ are
$$\bar{X} + 1.96(\sigma_{\bar{X}}) = 21.47 + 1.96(2.58) = 26.53$$
and
$$\bar{X} - 1.96(\sigma_{\bar{X}}) = 21.47 - 1.96(2.58) = 16.41$$

d. The 95 per cent confidence limits for μ are
$$\bar{X} + 2.14(S_{\bar{X}}) = 19.0 + 2.14(1.87) = 23.0$$
and
$$\bar{X} - 2.14(S_{\bar{X}}) = 19.0 - 2.14(1.87) = 15.0$$

e.
$$t = \frac{\bar{X} - \mu}{S_{\bar{X}}}$$
$$= \frac{21.47 - 20.00}{2.58}$$
$$= .57$$

$P(t = .57)$ is more than 5 per cent for 14df.

f.
$$t = \frac{\bar{X}_1 - \bar{X}_2}{S_{D\bar{x}}}$$
$$= \frac{21.47 - 19.00}{3.64}$$
$$= .68$$

$P(t = .68)$ is more than 5 per cent for 22df.

g. Same procedure as above.

h.

X_1	X_2	d	d^2
47	31	16	256
37	26	11	121
34	25	9	81
29	24	5	25
27	23	4	16
25	23	2	4
22	22	0	0
21	21	0	0
18	19	−1	1
17	16	1	1
13	16	−3	9
11	15	−4	16
8	14	−6	36
7	9	−2	4
6	1	5	25

$\Sigma = 322$ $\Sigma = 285$ $\Sigma = 37$ $\Sigma = 595$

$$S_d{}^2 = \frac{\Sigma d^2 - \dfrac{(\Sigma d)^2}{n}}{n}$$
$$= 33.58$$
$$S = 5.80$$
$$t = \frac{\bar{X}_1 - \bar{X}_2}{\dfrac{S_d}{\sqrt{n}}}$$
$$= \frac{2.47}{1.50}$$
$$= 1.65$$

$P(t = 1.65)$ is greater than 5 per cent for 14df.

IX

Analysis of Variance: I

The statistical procedures we considered in Chapter VIII dealt with tests of hypotheses about two populations. We saw that the probability ratio Z may be used to determine whether the differences between two events, $\bar{X}_1$ and $\bar{X}_2$, are large enough to warrant the rejection of a null-hypothesis. And we also saw that the probability ratio t, a special case of Z, may be used when the respective samples representing the populations being investigated are small enough to warrant an assumption of probable bias. This is true particularly when the sample size, n, is less than 30.

Both Z and t, however, are limited to providing an index of probability only when the differences between these *two* events, $\bar{X} - \mu$, or $\mu_1 - \mu_2$, are being considered: the event E in the $\dfrac{E}{E's}$ ratio can be the difference between *two* events only. Consequently, the Z and t ratios cannot be employed when more than two populations are under consideration.

It is a frequent occurrence in scientific investigation that more than two populations are being compared. Each of several groups in an experiment or study may be subject to a different treatment. The groups are then compared to determine whether the treatments were effective.

The basis for this kind of analysis is similar to that underlying procedures employed in testing hypotheses concerning two populations: the observations were originally obtained at random from a normally distributed population; then individuals are selected at random from the population and assigned to several subgroups, such as, for instance, a *control group* and several *treatment groups*. Individuals in the control group remain intact, while those in the treatment groups are subject to some sort of manipulation.

Most studies of this kind begin with one large sample divided into two or more smaller samples (subgroups). If the treatment is not effective, the

124

statistics of the respective treatment groups should not differ significantly from those of the control group. In turn, all subgroups should have distributions similar (statistically) to that of the population from which they were obtained.

If, however, the treatment group statistics differ significantly from those of the control group, a new assumption must be made: the distribution of the control group is similar to that of the population from which it came, but the treatment groups (or at least one of them) have a distribution like that of different populations, i.e. the population of individuals subject to that particular treatment.

The use of the term *treatment* in this context does not mean treatment of a medical nature. It must be interpreted in the more universal sense of its meaning. *Treatments* may consist of different methods of instruction of elementary mathematics to primary grade school children, or different amounts of food reward offered to rats performing a lever pressing task, etc. Treatment, then, refers to the particular condition imposed by the investigator on the members of a given sample or subgroup, i.e. the subjects comprising a treatment group.

At the beginning of this chapter, we noted that the use of Z and t applies only to the comparison $\dfrac{E}{E's}$ when E is the difference between two events. When more than two groups are employed in a study, the null-hypothesis would have to be

$$\bar{X}_1 - \bar{X}_2 - \bar{X}_3 \ldots - \bar{X}_k - \mu_1 - \mu_2 - \mu_3 \ldots - \mu_k = 0$$

for the comparison of the sample mean of k groups. There is, unfortunately, no table that provides the probability of the difference between k group means when k is greater than two. It is obvious that a different table would be needed for each possible value of k greater than two, i.e. $k = 3$, $k = 4$, etc.

The logical alternative to the above null-hypothesis is to compare all combinations of pairs of sample means. Such a test would result, if $k = 3$, for instance, in the following t tests:

Comparing groups I and II: $t = \dfrac{\bar{X}_1 - \bar{X}_2}{S_{\bar{X}_1 \bar{X}_2}}$

Comparing groups I and III: $t = \dfrac{\bar{X}_1 - \bar{X}_3}{S_{\bar{X}_1 \bar{X}_3}}$

Comparing groups II and III: $t = \dfrac{\bar{X}_2 - \bar{X}_3}{S_{\bar{X}_2 \bar{X}_3}}$

If $k = 4$, then there would be six such comparisons, and so on. The number of such tests required for k groups would simply be the number of *combinations* of k items taken two at a time.

The procedure of comparing all possible pairs of sample means is not statistically acceptable. This procedure is not valid for the following reason. Let us suppose that we are dealing with a population with mean, μ, and standard deviation, σ. The parametric values are not known. Should we wish to determine the sampling distribution of k means that could be obtained from this population, we need not actually generate the k sample means. We might simply obtain one reasonably large sample and make the assumption that the obtained sample mean and standard deviation are reasonable estimates of the population parametric values.

If the assumption is correct, then $\bar{X}$ is approximately $\bar{X}_{\bar{X}}$ and the latter is approximately equal to μ. Likewise, $S_{\bar{X}}$ is approximately equal to $\sigma_{\bar{X}}$ (for the particular df.). With the sample statistics, we can determine the α per cent confidence limits for μ. This procedure establishes *all* likely values of μ and *all* unlikely values of μ. There will be α per cent unlikely values of μ: these are rejected.

If, instead of establishing the confidence limits of μ, we had substituted all possible values of $\bar{X}$ in an unlimited number of t ratios, α per cent of these t ratios would likewise have been rejected. Thus, *the probability of obtaining significant t ratios increases with the number of t ratios formed.* Again, unfortunately, there is no table that tells us how the probability of obtaining significant t ratios changes as a function of the number of t ratios for samples of given *degrees of freedom.*

Because the increase in the number of t ratios increases the proportion of significant t ratios one is likely to obtain, the procedure of employing multiple simple t ratios for comparison of k groups is not recommended. However, there are other ways of determining whether there are significant differences between sample means of studies in which there are more than two groups.

When a particular treatment is effective, the difference between the statistics

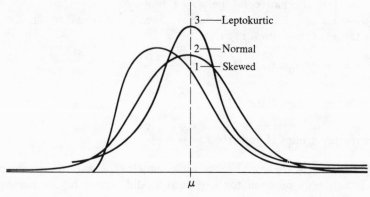

FIGURE 9.1

of the treatment group and the control group (sometimes differences between groups given different quantities of the treatment) are not limited to differences between group means. Not infrequently, the effectiveness of a treatment shows up as a significant difference in the sample variances. Under certain circumstances, it is possible that the only difference between the groups is a variances difference. Figure 9.1 shows three distributions with the same mean but with different variances.

One-variable Analysis of Variance

The analysis of variance is based, in part, on the contention that when several samples are obtained from a given population, each provides an independent estimate of the population variance barring biasing and limits imposed by differences between N and the respective sample df.

This type of analysis is, also, a special case of the $\dfrac{E}{E's}$ probability paradigm for determining the likelihood of certain events. The nature of the E and the distribution of alternatives, E's, will be clarified later on. To illustrate concepts and computations, in this type of analysis, the sample in Table 9.1 will be employed.

The one-variable analysis of variance is so designated because each observation in the classification table is made on a subject who belongs only to one treatment group. Thus, the one-variable analysis of variance model can be categorized as a treatments (columns) by subjects (rows) statistical design, which is often referred to as a *simple randomized* design.

	Treatment category			
	a	b	c	d
Subjects: 1	39	16	20	36
2	22	21	44	29
3	10	22	24	36
4	19	16	12	33
5	12	19	31	30
6	23	14	42	29
7	35	44	39	31
8	55	29	23	36
9	31	31	29	24
10	14	18	32	23

TABLE 9.1

Table 9.2 summarizes the manner in which the observations and statistics are labeled in the analysis of variance.

	Treatment categories (i)				
	a	b	c	d	
Subjects: 1	X_{a1}	X_{b1}	X_{c1}	X_{d1}	
2	X_{a2}	X_{b2}	X_{c2}	X_{d2}	
3	X_{a3}	X_{b3}	X_{c3}	X_{d3}	
4	X_{a4}	X_{b4}	X_{c4}	X_{d4}	
5	X_{a5}	X_{b5}	X_{c5}	X_{d5}	
6	.	.	.	.	
7	.	.	.	.	
8	.	.	.	.	
9	.	.	.	.	
10	X_{a10}	X_{b10}	X_{c10}	X_{d10}	
Sum:	T_{a+}	T_{b+}	T_{c+}	T_{d+}	T_{++}
Mean:	$\bar{X}_{a.}$	$\bar{X}_{b.}$	$\bar{X}_{c.}$	$\bar{X}_{d.}$	$\bar{X}_{..}$

TABLE 9.2

X_{ij} is any observation on subject j in category i. For example, X_{b3} is observation 3, in group b. In Table 9.2, X_{b3} is the observation 22.

The subscripts $+$ and $\cdot$ are used to indicate sums and means of columns and rows. They are placeholders, as it were. The first position indicates the column, and the second the row. For instance, T_{i+} is the sum of all observations in column i, and T_{+j} is the sum of all observations in the jth row. $\bar{X}_{i.}$ is the mean of the observations in column i and $\bar{X}_{.j}$ is the mean of the observations in the jth row.

$$T_{a+} = \Sigma X_a = \text{total (sum) of groups a.}$$

That is,

$$T_{a+} = X_{a1} + X_{a2} + X_{a3} \ldots + X_{an} = 39 + 22 + 10 \ldots + 14 = 260$$

The totals represent column sums and are referred to as category totals. Ordinarily, sample sums are indicated by the symbol Σ. However, by convention, in this kind of analysis the symbol T is used instead.

The category means are computed in the conventional way. The category totals are divided by the number of observations in the category:

$$\bar{X}_{a\cdot} = \frac{T_{a+}}{n_a}$$

T_{++} is the grand total and may be computed by summing all observations irrespective of category (ΣX_{ij}), or by obtaining the sum of the category totals:

$$T_{++} = T_{a+} + T_{b+} + T_{c+} + T_{d+}$$

The grand mean, $\bar{X}_{\cdot\cdot}$, can be obtained by dividing the sum of all observations irrespective of category by the number of observations in all groups combined:

$$\bar{X}_{\cdot\cdot} = \frac{\Sigma X_{ij}}{nk}$$

or

$$\bar{X}_{\cdot\cdot} = \frac{T_{++}}{nk}$$

where k is the number of categories.

For the data in Table 9.1, the grand total and grand mean are, respectively, $T_{++} = 1093$ and $\bar{X}_{\cdot\cdot} = \dfrac{1093}{40} = 27.32$.

The analysis of variance is based on what we know of the nature of sampling distributions of $\bar{X}$ and S^2, generated from a population from which k samples of size n have been constituted. Since, ordinarily, we do not know the parametric values, μ and σ^2, we can only test the significance of the difference between the sample statistics by reconstructing a hypothetical sampling distribution that would probably be generated by the unknown distribution from which the k samples presumably were obtained.

The variance of a sampling distribution of means can be estimated from the variance of the population, when it is known, for samples of size n:

$$\sigma_{\bar{X}}^2 = \frac{\sigma^2}{n}$$

It should be noted that the standard error can then be obtained by taking the square root of both terms:

$$\sigma_{\bar{X}} = \frac{\sigma}{\sqrt{n}}$$

When the population variance, σ^2, is not known, the variance of the sampling distribution can be estimated from the sample statistic S^2:

$$S_{\bar{X}}{}^2 = \frac{S^2}{n}$$

It should follow that if we were to take k samples and compute the variance of the means directly, then multiply this term by the sample size, n, we would have an estimate of the population variance, σ^2.

The Between-groups Sum-of-squares: $\Sigma x_b{}^2$

Let us suppose that we have obtained k samples of size n from a normally distributed population. These samples have been constituted by random sampling from the population and random assignment to the k groups. We can determine the mean of each of these samples, and get a sampling distribution of k means. See Table 9.3. The variance of this sampling distribution could be determined as follows:

$\bar{X}$	$(\bar{X} - \bar{X}_{\bar{X}})$	$(\bar{X} - \bar{X}_{\bar{X}})^2$
$\bar{X}_1$	$\bar{X}_1 - \bar{X}_{\bar{X}}$	$(\bar{X}_1 - \bar{X}_{\bar{X}})^2$
$\bar{X}_2$	$\bar{X}_2 - \bar{X}_{\bar{X}}$	$(\bar{X}_2 - \bar{X}_{\bar{X}})^2$
$\bar{X}_3$	$\bar{X}_3 - \bar{X}_{\bar{X}}$	$(\bar{X}_3 - \bar{X}_{\bar{X}})^2$
$\bar{X}_4$	$\bar{X}_4 - \bar{X}_{\bar{X}}$	$(\bar{X}_4 - \bar{X}_{\bar{X}})^2$
.	.	.
.	.	.
.	.	.
$\bar{X}_k$	$\bar{X}_k - \bar{X}_{\bar{X}}$	$(\bar{X}_k - \bar{X}_{\bar{X}})^2$
$\Sigma\bar{X}$	$\Sigma(\bar{X} - \bar{X}_{\bar{X}})$	$\Sigma(\bar{X} - \bar{X}_{\bar{X}})^2$

TABLE 9.3

The column labeled $\bar{X}$ is the sampling distribution of the k means. The sum of this column is the sum of the means, $\Sigma\bar{X}$, and the sum of the means divided by the number of means is the mean of the means, $\bar{X}_{\bar{X}}$, or the mean of the sampling distribution of means:

$$\bar{X}_{\bar{X}} = \frac{\Sigma\bar{X}}{k}$$

The next column is a distribution of the deviation of each mean from the mean of the means. The sum of this distribution is zero, just as it would be if it were the sum of the deviations of individual observations from the mean of a distribution. This column is labeled $(\bar{X} - \bar{X}_{\bar{X}})$.

In the next column, the deviation $(\bar{X} - \bar{X}_{\bar{X}})$ of the individual means from the means of the sampling distribution are squared: $(\bar{X} - \bar{X}_{\bar{X}})^2$. The sum of these squared deviations is the sum-of-squares for the sampling distribution of $\bar{X}$.

The sum-of-squares, divided by k, is the variance of the sampling distribution:

$$S_{\bar{X}}^2 = \frac{\Sigma(\bar{X} - \bar{X}_{\bar{X}})^2}{k}$$

The variance of the population from which the k samples were obtained can then be estimated as follows:

$$\sigma^2 \simeq n(S_{\bar{X}}^2)$$

There are alternative methods for computing $S_{\bar{X}}^2$. For instance, the calculator form for computing the sum-of-squares could have been employed:

$$\Sigma(\bar{X} - \bar{X}_{\bar{X}})^2 = \Sigma\bar{X}^2 - \frac{(\Sigma\bar{X})^2}{k}$$

When the term above is divided by k, it results in the same quantity.

The main point is, however, that the variance of a population can be estimated from the appropriate transformation of the variance of an empirically obtained sampling distribution.

This estimate is an essential component in the analysis of variance procedure.

Let us turn to the data in Table 9.1. Employing the appropriate symbol nomenclature, we can estimate the variance of the population from which the samples (k = 4, n = 10) were obtained.

The sum-of-squares of the sampling distribution is defined as

$$\Sigma x_M^2 = \Sigma(\bar{X}_{i\cdot} - \bar{X}_{\cdot\cdot})^2$$

This term is the sum of the squared deviations of the category (group or sample) means from the grand mean (previously identified as $\bar{X}_{\bar{X}}$, for illustrative purposes). The grand mean is

$$\bar{X}_{\cdot\cdot} = \frac{\bar{X}_{a\cdot} + \bar{X}_{b\cdot} + \bar{X}_{c\cdot} + \bar{X}_{d\cdot}}{k}$$

$$= \frac{26.0 + 23.0 + 29.6 + 30.7}{4}$$

$$= 27.325$$

X_a	x_a	x_a^2	X_b	x_b	x_b^2	X_c	x_c	x_c^2	X_d	x_d	x_d^2
39	+13	169	16	−7	49	20	+9.6	92.16	36	−5.3	28.09
22	−4	16	21	−2	4	44	−14.4	207.36	29	+1.7	2.89
10	−16	256	22	−1	1	24	+5.6	31.36	36	−5.3	28.09
19	−7	49	16	−7	49	12	+17.6	309.76	33′	−2.3	5.29
12	−14	169	19	−4	16	31	−1.4	1.96	30	+.7	.49
23	−3	9	14	−9	81	42	−12.4	153.76	29	+1.7	2.89
35	+9	81	44	+21	441	39	−9.4	88.36	31	−.3	.09
55	+29	841	29	+6	36	23	+6.6	43.56	36	−5.3	28.09
31	+5	25	31	+8	64	29	+.6	.36	24	+6.7	44.89
14	−12	144	18	−5	25	32	−2.4	5.76	23	+7.7	59.29
$\Sigma = 260$			230			296			307		
$\Sigma =$	0			0			0			0	
$\Sigma =$		1786			766			934.40			200.10
$\bar{X} = 26.0$			23.0			29.6			30.7		

TABLE 9.4

X_a	X_a^2	X_b	X_b^2	X_c	X_c^2	X_d	X_d^2
39	1521	16	256	20	400	36	1296
22	484	21	441	44	1936	29	841
10	100	22	484	24	576	36	1296
19	361	16	256	12	144	33	1089
12	144	19	361	31	961	30	900
23	529	14	196	42	1764	29	841
35	1225	44	1936	39	1521	31	961
55	3025	29	841	23	529	36	1296
31	961	31	961	29	841	24	576
14	196	18	324	32	1024	23	529
$\Sigma =$ 260		230		296		307	
$\Sigma =$	8546		6056		9696		9625

TABLE 9.5

Substituting our data, we obtain:

$$\Sigma x_M^2 = (26 - 27.325)^2 + (23 - 27.325)^2 +$$
$$(29.6 - 27.325)^2 + (30.7 - 27.325)^2$$
$$= (1.76) + (18.71) + (5.18) + (11.49)$$
$$= 37.03$$

We are interested, however, in estimating the population variance, σ^2, from which the k groups were drawn. Therefore, we transform the obtained sum-of-squares by multiplying it by n:

$$\Sigma x_b^2 = n(\Sigma x_M^2)$$
$$= 10(37.03)$$
$$= 370.30$$

This sum-of-squares term divided by the appropriate degrees of freedom $(k - 1)$ is an estimate of σ^2:

$$\sigma^2 \simeq n \frac{\Sigma(\bar{X}_i. - \bar{X}..)^2}{k - 1}$$
$$= \frac{370.30}{3}$$
$$= 123.43$$

It should be noted that the term Σx_b^2 divided by the appropriate df is known as the *mean square* (MS). In this case, it is MS_b.

The term MS_b can also be computed from the data in Table 9.5. The computational procedure is:

$$\Sigma x_b^2 = \Sigma \frac{T_{i+}^2}{n_i} - \frac{T_{++}^2}{nk}$$

$$= 30236.5 - 29866.22$$

$$= 370.28$$

$$MS_b = \frac{\Sigma x_b^2}{df}$$

$$= \frac{370.28}{3}$$

$$= 123.43$$

The Within-groups Sum-of-squares: Σx_w^2

The within-groups sum-of-squares, Σx_w^2, is defined as the sum of the sum-of-squares of the individual categories:

$$\Sigma x_w^2 = \Sigma\Sigma x_i^2 \quad \text{where } \Sigma x_{i\cdot}^2 = \Sigma(X_i - \bar{X}_{i\cdot})^2.$$
$$= \Sigma(X_a - \bar{X}_{a\cdot})^2 + \Sigma(X_b - \bar{X}_{b\cdot})^2 + \Sigma(X_c - \bar{X}_{c\cdot})^2 + \Sigma(X_d - \bar{X}_{d\cdot})^2$$

Substituting the data in Table 9.4, we obtain:

$$\Sigma x_w^2 = 1786 + 766 + 934.4 + 200.1$$
$$= 3686.5$$

The alternate calculator computation is:

$$\Sigma x_w^2 = \Sigma\Sigma X_{ij}^2 - \frac{\Sigma(T_{i+})^2}{n_i}$$

$$= 33923 - \frac{(260)^2 + (230)^2 + (296)^2 + (307)^2}{10}$$

$$= 33923 - 30236.5$$

$$= 3686.5$$

The within-groups sum-of-squares, divided by the appropriate $df\,(nk - k)$, is the pooled variance, S_p^2, which we encountered previously in Chapter III, in connection with a specific test concerning the difference between the mean of two groups. It can be defined as

$$S_p^2 = \frac{\Sigma(X_a - \bar{X}_{a\cdot})^2 + \Sigma(X_b - \bar{X}_{b\cdot})^2 + \Sigma(X_c - \bar{X}_{c\cdot})^2 + \Sigma(X_d - \bar{X}_{d\cdot})^2}{n_a + n_b + n_c + n_d - k}$$

This pooled variance is also an estimate of the population variance, σ^2:

$$\sigma^2 \simeq \frac{\Sigma x_w{}^2}{nk - k}$$

So far, we have obtained two independent estimates of the variance of the population from which the samples in the classification table were drawn. These two estimates of σ^2 are, respectively, MS_b, i.e. $\dfrac{\Sigma x_b{}^2}{k - 1}$, and MS_w, i.e. $\dfrac{\Sigma x_w{}^2}{nk - k}$.

The Total Sum-of-squares: $\Sigma x_t{}^2$

It is also possible to consider the observations in the classification table to constitute one sample of forty observations, rather than four samples of ten observations each. For such a sample, the total sum-of-squares divided by the appropriate df ($nk - 1$, or $N - 1$) is the variance of the total sample of forty observations:

$$\Sigma x_t{}^2 = \Sigma\Sigma(X_{ij} - \bar{X}..)^2$$

Substituting the data in Table 9.4, we obtain:

$$\Sigma x_t{}^2 = 4056.95$$

The calculator computation form may also be used, employing the data in Table 9.5:

$$\Sigma x_t{}^2 = \Sigma\Sigma X_{ij}{}^2 - \frac{T_{++}{}^2}{nk}$$

$$= 33923 - \frac{(1093)^2}{40}$$

$$= 33923.00 - 29866.22$$

$$= 4056.78$$

$$MS_t = \frac{\Sigma x_t{}^2}{N - 1}$$

$$= \frac{4056.77}{39}$$

$$= 104.01*$$

* This term is of no particular usefulness in the analysis of variance.

It is important to note that the total sum-of-squares in this design is the sum of the between- and within-groups sum-of-squares:

$$\Sigma x_t{}^2 = \Sigma x_b{}^2 + \Sigma x_w{}^2$$
$$4056.77 = 370.30 + 3686.5$$

The small differences are due to rounding-off errors.

Summary Table for One-variable Classification Analysis of Variance

The statistics obtained in the analysis of variance are generally reported in a standard summary table. The purpose of such a table is to indicate the source of the variance estimates. The format for this kind of table is shown in Table 9.6.

Summary table: Analysis of variance

Source	Sum-of-squares	df	Mean Square	F
Between	370.28	$k - 1 = 3$	123.43	
Within	3686.50	$nk - k = 36$	102.40	1.21
Total	4056.78	$N - 1 = 39$		

TABLE 9.6

The summary table for analysis of variance indicates the source of the variance estimates that are essential in a test of the significance of the differences between these variance estimates. This test is a ratio of the between groups mean square and the within-groups mean square. The probability that a given value of this ratio is significant at a given α level is based on percentiles of the F distribution for the particular df:

$$F = \frac{MS_b}{MS_w}$$
$$= \frac{123.42}{102.40}$$
$$= 1.21$$

If α is 5 per cent, and $df_1 = 3$, $df_2 = 32$, then the probability of $F = 1.21$ is greater than α. The probability that an F ratio (given the particular df) exceeds a value expected on the basis of chance 95 per cent of the time is given by the value in the F table, the *critical ratio*. See Table A.7.

Note that for $df_1 = 3$, $df_2 = 36$, the table value must be obtained by inter-polation. This interpolated value is 2.89. The value in the F table is the upper limit of $1 - \alpha$ per cent of the sampling distribution, F, for those df combinations. Any obtained value of an F ratio exceeding that value falls into the rejection region. Since our obtained value, 1.21, falls below the upper limit, we fail to reject the hypothesis that the F ratio is significantly greater than 1.00.

The failure to reject the obtained F ratio implies homogeneity of variances of the groups in the classification table.

Testing the Difference Between Category Means

A test of the significance of the differences between category means and vari-ances in the one-variable analysis of variance relies upon the parametric assumptions for the interpretation of the observed differences. The assump-tions have been somewhat modified for the specific hypotheses.

1. Categories have been obtained from the same population.
2. Observations in the categories are obtained at random, independently, from a normally distributed population.
3. The observations have been assigned, at random, to each of the cate-gories.
4. Category means are similar.
5. Category variances are similar.

If these criteria have been satisfied, then $\dfrac{MS_b}{MS_w}$ is distributed as F (for the particular df). If any of these conditions have *not* been met, the ratio may indicate significance of differences even if the different treatments are not differentially effective. This means, that F, like t and Z, may test a treatment effect, or it may reflect the extent to which there has been a failure to sample properly.

Consequently, it cannot be overemphasized that it is essential to determine how one might sample adequately, to make sure that decisions based on tests of significance are useful.

Testing the Significance of Differences Between Pairs of Sample Means

The F test is a *simultaneous* test. It specifies nothing about which particular treatment contributed to the significance of an obtained ratio. When the F ratio is significant, it is understood that the samples could not likely have come from the same population. It might still be desirable, however, to determine which group, if any, is significantly different from the others.

It is generally conceded that t tests, in conjunction with the analysis of variance, constitute questionable procedure. However, when it is necessary to determine whether a significant difference exists between pairs of means, and if the F ratio is not significant, the following procedure may be employed. First, the error variance must be determined. The error variance, that is, the square of the standard error of the differences between two samples $(S_{D_{\bar{X}}}^2)$, is, when the sample size is the same for the two samples:

$$S_{D_{\bar{X}}}^2 = MS_w \left(\frac{1}{n_1} + \frac{1}{n_2} \right)$$

The difference may then be tested:

$$t = \frac{\bar{X}_1 - \bar{X}_2}{S_{D_{\bar{X}}}}$$

For illustrative purposes, we shall test the difference between the mean of the first and second category in our table. These values are 26.0 and 23.0, respectively.

$$S_{D_{\bar{X}}}^2 = MS_w \left(\frac{1}{n_1} + \frac{1}{n_2} \right)$$
$$= 102.40 \left(\frac{1}{10} + \frac{1}{10} \right)$$
$$= 102.40(.1 + .1)$$
$$= 20.48$$
$$S_{D_{\bar{X}}} = \sqrt{S_{D_{\bar{X}}}^2}$$
$$= 4.53$$
$$t = \frac{26.0 - 23.0}{4.53}$$
$$= .66$$

The probability of $t = .66$, df $= n_1 + n_2 - 2 = 18$, is greater than .05, and we fail to reject the null-hypothesis.

When all category n's are equal, it is not necessary to compute t ratios for all possible pairs of means. A *critical difference index* may be calculated:

$$d = t_{1-\alpha} \left(\frac{\sqrt{2MS_w}}{n} \right)$$

where d is the difference between any pair of category means. To establish the appropriate value of t to be used, the upper limit of the t distribution for a particular level of significance is determined for the df corresponding to MS_w.

Testing the Significance of Differences Between Category Means when the Variances Are Not Equal

The following test has been suggested (by Cochran and Cox, see Bibliography) for testing the significance of the difference between category means when the F test suggests heterogeneity of variances:

$$t' = \frac{(S_{\bar{X}_1}^2)t_1 + (S_{\bar{X}_2}^2)t_2}{S_{\bar{X}_1}^2 + S_{\bar{X}_2}^2}$$

The ratio

$$\frac{\bar{X}_1 - \bar{X}_2}{S_{D\bar{X}}}$$

is significant if it is greater than t'. The values of t_1 and t_2 are determined from the t tables for the df associated with the particular $S_{\bar{X}}$, and

$$S_{D\bar{X}} = \sqrt{\frac{\Sigma x_1^2}{n_1(n_1 - 1)} + \frac{\Sigma x_2^2}{n_2(n_2 - 2)}}$$

Testing for Excessive Variability Among Category Means

The following procedure allows an estimate of the variability of the category means about the grand mean. The test is

$$F = \frac{\dfrac{\Sigma(\bar{X} - \bar{X}_{\bar{X}})^2}{k - 1}}{S_{\bar{X}}^2}$$

where $\bar{X}$ is any mean in the classification table, $\bar{X}_{\bar{X}}$ is the grand mean ($\bar{X}..$), $k - 1$ is the df associated with between-groups variance, and $S_{\bar{X}}^2 = \dfrac{MS_w}{\sqrt{df}}$.

For illustrative purposes, substituting our data, we obtain:

$$F = \frac{\dfrac{(26.00 - 27.32)^2 + (23.00 - 27.32)^2 + (29.60 - 27.32)^2 + (30.70 - 27.32)^2}{3}}{16.41}$$

$$= \frac{12.34}{16.41}$$

$$= .76$$

If α is 5 per cent, we reject the hypothesis that there is excessive variability among category means; i.e. we fail to reject the null-hypothesis.

Testing the Significance of the Differences Between Category Variances

A significant difference may be obtained, in the analysis of variance, if one of the category variances is considerably greater than the others.

The statistic C (in the context of Cochran's test) may be used to determine whether the differences between the largest S^2 and the others is significant:

$$C = \frac{\text{largest } S^2}{\Sigma S^2}$$

That is, C is the ratio of the largest variance to the sum of the variance of the categories. The probability of C can be determined from Table A.6. Critical values of C are given for various $1 - \alpha$ per cent points of the sampling distributions, C, for k, in the numerator and n-1 in the denominator. (k is the number of categories in the term ΣS^2). Substituting our data, we obtain:

$$S^2 = \Sigma X^2 - \frac{(\Sigma X)^2}{n} \div n$$

$$S_a^2 = 8546 - \frac{(67600)}{10} \div 10 = 178.60$$

$$S_b^2 = 6056 - \frac{(52900)}{10} \div 10 = 76.60$$

$$S_c^2 = 9696 - \frac{(87616)}{10} \div 10 = 93.44$$

$$S_d^2 = 9625 - \frac{(94249)}{10} \div 10 = 20.01$$

$$C = \frac{\text{largest } S^2}{\Sigma S^2}$$

$$= \frac{178.60}{368.65}$$

$$= .484$$

The exact table value for 9 and 27 df in Table A.6 must be obtained by interpolation. Since it does not make much difference, given the obtained value of C, 9 by 30 df is close enough. That table value is .0958 for the 95 per cent confidence level. Since the obtained C values is greater than the table value, we conclude that S_a^2 is significantly different from the other S^2 in the classification table.

Problems

1. Given fifty observations, k = 5, n = 10:

Group a	Group b	Group c	Group d	Group e
17	37	9	35	11
11	8	22	20	27
18	25	24	29	14
27	7	15	10	19
6	47	16	22	32
22	23	31	21	15
34	1	26	11	33
13	21	14	35	28
21	25	16	1	26
29	23	19	16	30

a. Test the hypothesis that the five samples above are probably random samples from the same normally distributed population.
b. Test the hypothesis that the difference between sample means is not significant at the 99 per cent level of confidence.
c. Test the hypothesis that the sample variances are not significantly different at the 95 per cent level of confidence.

Answers

1.

	Group a		Group b		Group c		Group d		Group e	
	X	X^2	X	X^2	X	X^2	X	X^2	X	X^2
	17	289	37	1369	9	81	35	1225	11	121
	11	121	8	64	22	484	20	400	27	729
	18	324	25	625	24	576	29	841	14	196
	27	729	7	49	15	225	10	100	19	361
	6	36	47	2209	16	256	22	484	32	1024
	22	484	23	529	31	961	21	441	15	225
	34	1156	1	1	26	676	11	121	33	1089
	13	169	21	441	14	196	35	1225	28	784
	21	441	25	625	16	256	1	1	26	676
	29	841	23	529	19	361	16	256	30	900
$\Sigma =$	198	4950	217	6441	192	4072	200	5094	235	6105

$$\Sigma X_a = 198 \qquad (\Sigma X_a)^2 = 39204 \qquad \bar{X}_a = 19.8 \qquad \Sigma X_a^2 = 4950$$

$$\Sigma X_b = 217 \qquad (\Sigma X_b)^2 = 47089 \qquad \bar{X}_b = 21.7 \qquad \Sigma X_b^2 = 6441$$

$$\Sigma X_c = 192 \qquad (\Sigma X_c)^2 = 36864 \qquad \bar{X}_c = 19.2 \qquad \Sigma X_c^2 = 4072$$

$$\Sigma X_d = 200 \qquad (\Sigma X_d)^2 = 40000 \qquad \bar{X}_d = 20.0 \qquad \Sigma X_d^2 = 5094$$

$$\Sigma X_e = 235 \qquad (\Sigma X_e)^2 = 55225 \qquad \bar{X}_e = 23.5 \qquad \Sigma X_e^2 = 6105$$

and

$$\Sigma x^2 = \Sigma X^2 - \frac{(\Sigma X)^2}{n} \quad \text{and} \quad S^2 = \frac{\Sigma x^2}{n - 1}$$

$$\Sigma x_a^2 = 4590 - 3920.4 = 669.6 \qquad S_a^2 = 74.40$$

$$\Sigma x_b^2 = 6441 - 4708.9 = 1732.1 \qquad S_b^2 = 192.46$$

$$\Sigma x_c^2 = 4072 - 3686.4 = 385.6 \qquad S_c^2 = 42.84$$

$$\Sigma x_d^2 = 5094 - 4000.0 = 1094.0 \qquad S_d^2 = 121.56$$

$$\Sigma x_e^2 = 6105 - 5522.5 = 582.5 \qquad S_e^2 = 64.72$$

The between-groups sum-of-squares:

$$\Sigma x_b^2 = n\Sigma(\bar{X} - \bar{X}..)^2$$
$$= 10(12.30)$$
$$= 123.00$$

The within-groups sum-of-squares:

$$\Sigma x_w^2 = \Sigma x_a^2 + \Sigma x_b^2 + \Sigma x_c^2 + \Sigma x_d^2 + \Sigma x_e^2$$
$$= 669.6 + 1732.1 + 385.6 + 1094.0 + 582.5$$
$$= 4463.80$$

Summary Table:

Source	Sum-of-squares	df	Mean square	F ratio
Between	123.00	4	30.75	$\dfrac{30.75}{99.16} = .31$
Within	4463.80	45	99.16	
Total	4586.80	49		

1. a. The probability of $F = .31$, $df_1 = 4$, $df_2 = 45$, is greater than 5 per cent, and we fail to reject the null-hypothesis.

   b. One over-all test of differences between category means might be a test of "excessive variability":

$$F = \frac{\dfrac{\Sigma(\bar{X} - \bar{X}_{\bar{X}})^2}{k - 1}}{S_{\bar{X}}^2}$$

where

$$S_{\bar{X}}^2 = \frac{MS_w}{\sqrt{df}}$$

$$= \frac{99.16}{6.41}$$

$$= 14.78$$

   and

$$F = \frac{\dfrac{(19.8 - 20.84)^2 + (21.7 - 20.84)^2 + (19.2 - 20.84)^2 + (20.0 - 20.84)^2 + (23.5 - 20.84)^2}{4}}{14.78}$$

$$= \frac{\dfrac{1.08 + .74 + 2.69 + .71 + 7.08}{4}}{14.78}$$

$$= .21$$

$F = .21$, $df_1 = 4$, $df_2 = 45$, is not significant at the 1 per cent level.

   c. Testing a hypothesis about the sample variances may be accomplished using Cochran's test.

$$C = \frac{\text{largest } S^2}{\Sigma S^2}$$

$$= \frac{192.46}{459.98}$$

$$= .388$$

$C = .388$, $df_1 = 9$, $df_2 = 45$, exceeds the upper 95 per cent limit of the sampling distribution, and we reject the null-hypothesis.

X

Analysis of Variance: II

In Chapter IX we presented a relatively simple computational procedure for the comparison of two independently obtained estimates of population variance. In that particular design, you will recall, each group, sample, or category consisted of one treatment classification. That is, each observation in the classification table is identified as having been obtained from an individual belonging to one treatment group only. The test of significance, based on the F ratio, is related to the extent to which the population variance, estimated from the empirically obtained sampling distribution

$$(\bar{X}_{a.} - \bar{X}_{..}), \quad (\bar{X}_{b.} - \bar{X}_{..}), \quad (\bar{X}_{c.} - \bar{X}_{..}) \ldots, \quad (\bar{X}_{k.} - \bar{X}_{..})$$

differs from an estimate of the population variance based on a pooling of the sum-of-squares of all categories.

In terms of the $\dfrac{E}{E's}$ probability paradigm, the numerator of the F ratio consists of a particular event: MS_b. Actually, MS_b should be approximately equal to σ^2, if treatment has no effect and parametric assumptions have been satisfied. The denominator of the ratio is MS_w, likewise approximately equal to σ^2 under specifiable conditions. The relationship between Z and t has already been explored at length. Let us examine the relationship between t and F in the comparison of two groups:

$$t = \frac{\bar{X}_1 - \bar{X}_2}{S_p \sqrt{n}}$$

$$F = \frac{n \dfrac{\Sigma(\bar{X} - \bar{X}_{\bar{x}})^2}{k - 1}}{S_p{}^2}$$

In the case of t, above, the event in question is the difference between two events: $\bar{X}_1$, and $\bar{X}_2$, and the distribution in question is a sampling distribution of such differences, since $S_p\sqrt{n}$ is an estimate of σ_p. In the case of F, the event examined is the variance of the observed sampling distribution transformed by multiplication by n, and the distribution with which it is being compared is the distribution (in terms of its variance, σ^2) of the population from which the sampling distribution theoretically was obtained. Close examination of the two formulas for t and F, above, results in the conclusion that F is approximately equal to t^2; the comparisons are $\dfrac{E}{E's}$, versus $\dfrac{E^2}{E's^2}$, respectively.

The one-variable classification analysis of variance has certain shortcomings. Under many conditions, it is desirable to assign an individual to more than one treatment group and to observe the effects of the several treatments on the subject simultaneously. With the one-variable analysis, r replications would be required with the k groups.

Let us suppose, hypothetically, that we wish to determine the effects of three reinforcement quantities, a, b, and c, on maze performance of rats in three different types of mazes: A, B, and C. We wish to examine the populations of observations that might be expected when subjects are classified by two characteristics. Such a procedure makes it possible to perform two studies simultaneously. For each variable, there are a number of different levels. If there are an equal number of observations in each level for each variable, the analysis is known as a *factorial* design.

Two-variable Classification Analysis of Variance (Single Observation)

The analysis outlined below is a factorial analysis with one observation for each level. The observations are designated with respect to the same scheme as the one above.

Basically, this classification table tells us that we are dealing with two separate approaches to an estimate of σ^2 based on the between-groups sum-of-squares: one for columns and one for rows. In the one-variable design the variable was totally identified with columns. In this design, we can turn the classification table 90° clockwise and we still have a design consisting of observations in columns. Thus the two-variable analysis is the simple superimposition of two one-variable classification tables.

The two classifications are variable I, groups A, B, C, and D, and variable II, groups a, b, and c, as shown in Figure 10.1. To consider the relationship between the two variables we shall examine how the sum-of-squares of the

Variable I

		A	B	C	D	Total	Mean
	a	X_{A_a}	X_{B_a}	X_{C_a}	X_{D_a}	T_{+a}	$\bar{X}_{\cdot a}$
Variable II	b	X_{A_b}	X_{B_b}	X_{C_b}	X_{D_b}	T_{+b}	$\bar{X}_{\cdot b}$
	c	X_{A_c}	X_{B_c}	X_{C_c}	X_{D_c}	T_{+c}	$\bar{X}_{\cdot c}$
Total		T_{A+}	T_{B+}	T_{C+}	T_{D+}	T_{++}	
Mean		$\bar{X}_{A\cdot}$	$\bar{X}_{B\cdot}$	$\bar{X}_{C\cdot}$	$\bar{X}_{D\cdot}$		$\bar{X}_{\cdot\cdot}$

TABLE 10.1

"column" categories and "row" categories contribute to the total sum-of-squares: all between- and within-groups sum-of-squares components should, when added, equal the total sum-of-squares.

To illustrate computational procedures, the data in Figure 10.2 will be employed. There are four groups in the variable I classification and three groups in the variable II classification.

Variable I

		A	B	C	D
	a	30	17	32	36
Variable II	b	26	14	29	24
	c	40	32	15	31

	A		B		C		D		T	$\bar{X}$	T^2
	X	X^2	X	X^2	X	X^2	X	X^2			
a	30	900	17	289	32	1024	36	1296	115	38.33	3509
b	26	676	14	196	29	841	24	576	93	31.00	2289
c	40	1600	32	1024	15	225	31	961	118	39.33	3810
T	96		63		76		91		326		
$\bar{X}$	32.00		21.00		25.33		30.33			27.17	
T^2		3176		1509		2090		2833			9608

TABLE 10.2

The Total Sum-of-squares: Σx_t^2

The procedure used in computing the total sum-of-squares is the same as that used in the simple (one-variable) analysis:

$$\Sigma x_t^2 = \Sigma\Sigma X_{ij}^2 - \frac{T_{++}^2}{rc}$$

where r is the number of categories in the rows, and c is the number of categories in the columns.

$$\begin{aligned}
\Sigma x_t^2 &= \Sigma\Sigma X_{ij}^2 - \frac{T_{++}^2}{rc} \\[4pt]
&= 9608 - \frac{(326)^2}{12} \\[4pt]
&= 9608 - 8856.33 \\[4pt]
&= 751.67
\end{aligned}$$

The total sum-of-squares has $rc - 1$ df. In this case, df $= 11$.

The Sum-of-squares for Column Means: $\Sigma x_{b_c}^2$

The two-variable classification table can be considered to be as previously indicated, two one-variable tables; the term $\Sigma x_{b_c}^2$ is obtained by computations that are based on a consideration of the variable I classification, shown below:

		A	B	C	D	
			Variable I			
Subject:	1	X_{A1}	X_{B1}	X_{C1}	X_{D1}	
	2	X_{A2}	X_{B2}	X_{C2}	X_{D2}	
	3	X_{A3}	X_{B3}	X_{C3}	X_{D3}	
		T_{A+}	T_{B+}	T_{C+}	T_{D+}	T_{++}
		$\bar{X}_{A\cdot}$	$\bar{X}_{B\cdot}$	$\bar{X}_{C\cdot}$	$\bar{X}_{D\cdot}$	$\bar{X}_{\cdot\cdot}$

Then the term is computed as before:

$$\begin{aligned}
\Sigma x_{b_c}^2 &= \Sigma \frac{T_{i+}^2}{n_i} - \frac{T_{++}^2}{rc} \\[4pt]
&= \frac{(96)^2}{3} + \frac{(63)^2}{3} + \frac{(76)^2}{3} + \frac{(91)^2}{3} - \frac{(326)^2}{12} \\[4pt]
&= 9080.67 - 8856.33 \\[4pt]
&= 224.34
\end{aligned}$$

The sum-of-squares for column means has $c - 1$ df. In this case, df $= 3$.

The Sum-of-squares for Row Means: $\Sigma x_{b_r}^2$

The sum-of-squares for row means, like the sum-of-squares for column means, may be used as the basis for an estimate of the population variance, σ^2, based on the distribution of variable II.

		Subjects					
		1	2	3	4	Total	Mean
	a	X_{1a}	X_{2a}	X_{3a}	X_{4a}	T_{+a}	$\bar{X}_{\cdot a}$
Variable II	b	X_{1b}	X_{2b}	X_{3b}	X_{4b}	T_{+b}	$\bar{X}_{\cdot b}$
	c	X_{1c}	X_{2c}	X_{3c}	X_{4c}	T_{+c}	$\bar{X}_{\cdot c}$
						T_{++}	
							$\bar{X}_{\cdot\cdot}$

It is computed as follows:

$$\Sigma x_{b_r}^2 = \frac{\Sigma T_{+j}^2}{n_j} - \frac{T_{++}^2}{rc}$$

$$= \frac{(115)^2}{4} + \frac{(93)^2}{4} + \frac{(118)^2}{4} - \frac{(326)^2}{12}$$

$$= 8949.50 - 8856.33$$

$$= 93.17$$

The sum-of-squares for row means has $r - 1$ df. In this case, df $= 2$.

The Residual Sum-of-squares: Σx_{res}^2

The residual sum-of-squares is obtained, literally, as a residual: it is the total sum-of-squares less the between-columns and between-rows sum-of-squares.

$$\Sigma x_{res}^2 = \Sigma x_t^2 - (\Sigma x_{b_c}^2 + \Sigma x_{b_r}^2)$$

Substituting the data we have computed so far:

$$\Sigma x_{res}^2 = 751.67 - (224.34 + 93.17)$$

$$= 434.16$$

The Summary Table

The sum-of-squares terms obtained above can be summarized in a table listing their source, the df for each term, and the MS. The MS's can then be used to form ratios testing specific hypotheses about column and row means. The table is given below.

Summary Table

Source	Sum-of-squares	df	Mean square
Row means	93.17	$r - 1 = 2$	46.58
Column means	224.34	$c - 1 = 3$	74.78
Residual	434.16	$(c - 1)(r - 1) = 6$	72.36
Total	751.67	$rc - 1 = 11$	

Testing the Effect of Variable I

Variable I consists of treatments whose effect, if any, should be manifest as a significant difference between column means and column variances. To test for the over-all significance of these differences, an F ratio is formed between the mean square for columns and the residual mean square:

$$F = \frac{MS_{b_c}}{MS_{res}}$$
$$= \frac{74.78}{72.36}$$
$$= 1.03$$

The probability of $F = 1.03$, $df_1 = 3$, $df_2 = 6$, is greater than 5 per cent and we fail to reject the hypothesis that such a ratio could not have been obtained on the basis of chance alone within the limits of the criterion for chance. We assume, therefore, that the variable I treatments were not differentially effective.

Testing the Effect of Variable II

The second variable consists of treatments whose effect should result in row differences. The significance of these differences can be tested with the ratio of the rows mean square and the residual mean square:

$$F = \frac{MS_{b_r}}{MS_{res}}$$
$$= \frac{46.58}{72.36}$$
$$= .64$$

Since the probability of $F = .64$, $df_1 = 2$, $df_2 = 6$ is greater than 5 per cent, we fail to reject the hypothesis that the differences between the rows is significant and conclude that the second variable also had no particular effectiveness in that the treatments did not result in significantly different group distributions.

Two-variable Classification Analysis of Variance (Multiple Observations)

The precision and reliability of any experimental investigation is considerably increased when observations of the effect of treatments is replicated: precision increases, within limits, with the repetition of observations.

The analysis of variance described below is similar to the one previously discussed, but, in addition, it takes into account repeated observations within each of the rc cells, in which there was previously only one observation.

The nature of this type of analysis and the comparisons that constitute tests of significance hinge about the residual component of variance, which is sometimes referred to as the "interaction." To understand this residual, and its relationship to column and row effects, let us examine the two one-variable tables in Table 10.3.

Variable I

		A	B	C	T	X̄
	a	1	2	3	6	2
Variable II	b	2	3	1	6	2
	c	3	1	2	6	2
	T	6	6	6	$T_{++} = 18$	
	X̄	2	2	2		$\bar{X}.. = 2$

TABLE 10.3(a)

Variable I

		A	B	C	T	X̄
	a	1	2	3	6	2
Variable II	b	1	2	3	6	2
	c	3	1	2	6	2
	T	5	5	8	$T_{++} = 18$	
	X̄	1.67	1.67	2.66		$\bar{X}.. = 2$

TABLE 10.3(b)

For the sake of simplicity, the two classification tables, Tables 10.3(a) and 10.3(b), have one observation per rc cell. The appropriate sum-of-squares terms are computed and summarized below.

For classification table A:

$$\Sigma x_t^2 = \Sigma\Sigma X_{ij}^2 - \frac{T_{++}^2}{nrc} = 42 - \frac{(18)^2}{9} = 6$$

$$\Sigma x_{b_c}^2 = \frac{\Sigma T_{i+}^2}{n_i} - \frac{T_{++}^2}{nrc} = \frac{(6)^2 + (6)^2 + (6)^2}{3} - 36 = 0$$

$$\Sigma x_{b_r}^2 = \frac{\Sigma T_{+j}^2}{n_j} - \frac{T_{++}^2}{nrc} = \frac{(6)^2 + (6)^2 + (6)^2}{3} - 36 = 0$$

$$\Sigma x_{res}^2 = \Sigma x_t^2 - \Sigma x_{b_c}^2 - \Sigma x_{b_r}^2 = 6 - 0 - 0 = 6$$

Summary Table

Source	Sum-of-squares	df	Mean square
r	0	2	0
c	0	2	0
Residual	6	4	1.5
Total	6	8	

For classification table B:

$$\Sigma x_t^2 = \Sigma\Sigma X_{ij}^2 - \frac{T_{++}^2}{nrc} = 42 - 36 = 6 \text{ (same as above)}$$

$$\Sigma x_{b_c}^2 = \frac{\Sigma T_{i+}^2}{n_i} - \frac{T_{++}^2}{nrc} = \frac{(5)^2 + (5)^2 + (8)^2}{3} - 36 = 38 - 36 = 2$$

$$\Sigma x_{b_r}^2 = \frac{\Sigma T_{+j}^2}{n_j} - \frac{T_{++}^2}{nrc} = \frac{(6)^2 + (6)^2 + (6)^2}{3} - 36 = 0$$

$$\Sigma x_{res}^2 = \Sigma x_t^2 - \Sigma x_{b_c}^2 - \Sigma x_{b_r}^2 = 6 - 2 - 0 = 4$$

Summary Table

Source	Sum-of-squares	df	Mean square
r	0	2	0
c	2	2	1
Residual	4	4	1
Total	6	8	

There is no point in performing the F tests, but the differences between the two classification tables clearly show why the residual sum-of-squares reflects the difference in column and row distributions. In the first classification table, 10.3(a), all marginal totals and means are identical. Thus, the effect of variable I and II are equally distributed among the *rc* cells.

In the second classification table, 10.3(b), the effects of variable II are equally distributed among row totals and means, but the effects of variable I in group C can be seen to differ from those in groups A and B. This difference is reflected in the column totals and means. And the differences in column and row distributions are further seen in the summary table in terms of the corresponding MS terms.

Thus, whenever the *rc* cell observations in the classification table depart from the random value distribution, the residual sum-of-squares will be affected.

It can be seen here that in analyses of this type the residual sum-of-squares depends on the contribution of the rows and columns sum-of-squares. The residual sum-of-squares reflects differences between the distribution of the columns and rows.

To illustrate the analysis of variance with repeated observations, let us examine the following data. The analysis is based upon the terms computed below.

The Total Sum-of-squares: Σx_t^2

$$\Sigma x_t^2 = \Sigma\Sigma X_{ij}^2 - \frac{T_{++}^2}{rcn}$$

Substituting the data in Table 10.4, we obtain:

$$\Sigma x_t^2 = 5993 + 4470 + 2956 + 3089 + 2106 + 4634 + 4410 + 2945$$
$$+ 3046 - \frac{(1151)^2}{45}$$

$$= 33649 - \frac{1324801}{45}$$

$$= 4208.98$$

The total sum-of-squares in this analysis is similar to the total sum-of-squares in previous analysis of variance designs. In this particular type of analysis, Σx_t^2 has $nk - 1$ df.

The Sum-of-squares for Category Means: Σx_b^2

This term is computed in the same way as Σx_b^2 encountered in connection with the one-variable analysis. The estimate of variance that can be obtained from this term is based on the sampling distribution of k categories; k in this case is 9. Thus, this term is based on an analysis of nine independently constituted samples, ignoring variable classification.

$$\Sigma x_b{}^2 = \Sigma \frac{T_{i+}{}^2}{n_i} - \frac{T_{++}{}^2}{rcn}$$

Substituting the data:

$$\Sigma x_b{}^2 = \frac{(171)^2}{5} + \frac{(138)^2}{5} + \frac{(112)^2}{5} + \frac{(117)^2}{5} + \frac{(96)^2}{5} + \frac{(142)^2}{5} + \frac{(146)^2}{5}$$

$$+ \frac{(111)^2}{5} + \frac{(118)^2}{5} - 29440.02$$

$$= \frac{151459}{5} - 29440.02$$

$$= 851.78$$

This sum-of-squares term, $\Sigma x_b{}^2$, has $rc - 1$ df.

The Sum-of-squares for Column Means: $\Sigma x_{bc}{}^2$

In the two-variable analysis of variance, single observation, the between-means sum-of-squares was not a necessary computational step. Since each cell contained only one observation, there was no within-cell variance. Consequently, only column and row effects were relevant. In this particular case, since there is an actual distribution in each rc cell, the variance within that rc cell distribution is taken into account in addition to the row and column variance about the grand mean. Theoretically, the between, within, and residual sum-of-squares are components of the category means sum-of-squares.

In this design, the sum-of-squares for column means is based upon differences between columns, ignoring rows. It is therefore based on the effect of variable I, irrespective of the effect of variable II.

$$\Sigma x_{bc}{}^2 = \Sigma\Sigma \frac{T_{i+}{}^2}{\Sigma n_i} - \frac{T_{++}{}^2}{rcn}$$

Substituting the data in Table 10.4,

$$\Sigma x_{bc}{}^2 = \frac{(421)^2}{15} + \frac{(355)^2}{15} + \frac{(375)^2}{15} - 29440.02$$

$$= 29592.73 - 29440.02$$

$$= 152.71$$

The sum-of-squares for column means has $c - 1$ df.

The Sum-of-squares for Row Means: Σx_{br}^2

The sum-of-squares for row means is based upon row or variable II differences, ignoring column or variable I effects.

$$\Sigma x_{b_r}^2 = \Sigma\Sigma \frac{T_{+j}^2}{\Sigma n_j} - \frac{T_{++}^2}{rcn}$$

Substituting the data, we obtain:

$$\Sigma x_{b_r}^2 = \frac{(434)^2}{15} + \frac{(345)^2}{15} + \frac{(372)^2}{15} - 29440.02$$

$$= \frac{445765}{15} - 29440.02$$

$$= 277.65$$

The sum-of-squares for row means has $r - 1$ df.

The Residual Sum-of-squares: Σx_{int}^2

The residual sum-of-squares, as in the previous design, is that sum-of-squares component which is unaccounted for by either row or column differences but which is the result of over-all rc cell differences. The residual sum-of-squares is obtained as before:

Variable I

		A	B	C
		30	24	33
		26	12	29
	a	40	36	20
		36	28	28
		39	17	36
		34	14	12
		37	18	22
Variable II	b	10	32	27
		18	21	12
		39	11	38
		21	27	32
		12	27	29
	c	15	48	11
		25	26	24
		39	14	22

Variable I

		A		B		C		
		X	X²	X	X²	X	X²	ΣT
		30	900	24	576	33	1089	
		26	676	12	144	29	841	
	a	40	1600	36	1296	20	400	
		36	1296	28	784	28	784	
		39	1521	17	289	36	1296	
	T =	171	5993	117	3089	146	4410	434
		34	1156	14	196	12	144	
		37	1369	18	324	22	484	
Variable II	b	10	100	32	1024	27	729	
		18	324	21	441	12	144	
		39	1521	11	121	38	1444	
	T =	138	4470	96	2106	111	2945	345
		21	441	27	729	32	1024	
		12	144	27	729	29	841	
	c	15	225	48	2304	11	121	
		25	625	26	676	24	576	
		39	1521	14	196	22	484	
	T =	112	2956	142	4634	118	3046	372
	ΣT =	421		355		375		1151

$$\Sigma x_{\text{int}}^2 = \Sigma x_{\text{b}}^2 - \Sigma x_{\text{b}_\text{c}}^2 - \Sigma x_{\text{b}_\text{r}}^2$$

From our data we may substitute the following computed values:

$$\Sigma x_{\text{int}}^2 = 851.78 - 152.71 - 277.65$$
$$= 421.42$$

In this type of design, the residual sum-of-squares is called the *interaction* sum-of-squares and has $(c - 1)(r - 1)$ df.

The Within-groups Sum-of-squares: Σx_w^2

The within-groups sum-of-squares can be obtained by subtraction:

$$\Sigma x_w^2 = \Sigma x_t^2 - \Sigma x_b^2$$

Substituting the computed values,

$$\Sigma x_w^2 = 4208.98 - 851.78$$
$$= 3357.20$$

The within-groups sum-of-squares can also be computed directly, adapting the formula appropriate in the single observation design:

$$\Sigma x_w^2 = \Sigma\Sigma X_{ij}^2 - \Sigma\Sigma \frac{T_{i+}^2}{n_i}$$
$$= 33649.00 - 30291.80$$
$$= 3357.20$$

The within-groups sum-of-squares has $nk - rc$ df.

Analysis of Variance Summary Table

The computed sum-of-squares components are summarized in a table indicating source, df, and mean square.

Summary Table

Source	Sum-of-squares	df	Mean square
Rows	277.65	2	138.82
Columns	152.71	2	76.36
Interaction	421.42	4	105.36
Subtotal (means)	851.78	8	106.47
Within	3357.20	36	93.26
Total	4208.98	44	

Testing the Significance of Differences Between Populations

The mean square terms necessary for testing specific hypotheses about the distributions in the classification table are summarized in the table above. Note that the first portion of the table, including rows, columns, and interaction mean squares, provides data for testing variable I and variable II effects. This is similar to the classification table previously encountered in connection with the two-variable classification, single observation.

Testing Significance of Differences Between Variable I Categories

To determine whether there are significant differences between the variable I treatment groups, we form a ratio between the columns mean square and the within-groups mean square:

$$F = \frac{MS_{b_c}}{MS_w}$$

Substituting the respective MS terms in the summary table, we obtain

$$F = \frac{76.36}{93.26}$$

$$= .82$$

The probability of $F = .82$, $df_1 = 2$, $df_2 = 36$, is greater than 5 per cent. Thus we conclude that there are no significant differences between the variable I treatment groups.

Testing Significance of Differences Between Variable II Categories

The significance of differences attributable to the variable II treatments can be determined by forming a ratio between the rows mean square and the within-groups mean square:

$$F = \frac{MS_{b_r}}{MS_w}$$

Substituting the appropriate terms from the summary table,

$$F = \frac{138.82}{93.26}$$

$$= 1.49$$

$F = 1.49$, $df_1 = 2$, $df_2 = 36$, is greater than 5 per cent, and we fail to reject the null-hypothesis.

Testing the Significance of Interaction

To test for significant interaction, we form a ratio between the interaction mean square and the within-groups mean square:

$$F = \frac{MS_{int}}{MS_w}$$

Substituting the respective terms from the summary table,

$$F = \frac{105.36}{93.26}$$

$$= 1.13$$

The probability that $F = 1.13$, $df_1 = 4$, $df_2 = 36$, is greater than 5 per cent and we conclude that interaction is not significant.

Using the data in the summary table, we have tested the significance of the differential effects of the variable I and II treatments, and the significance of interaction. None of these was found to be significant at the 5 per cent level.

The test for the variable I and variable II treatments is reasonably clear. The test for interaction in this case is not quite so clear. Interaction can be understood in the following terms: MS_{int} depends on the variability of the rc cell means. If the interaction test results in an F not significant at α per cent level, it may be concluded that rc cell fluctuations do not exceed what may be expected on the basis of chance.

Problems

1. Given the following observations, test the significance of the differences between the variable I and variable II treatments:

		Variable I		
		A	B	C
	a	9	8	9
Variable II	b	3	5	6
	c	6	6	4

2. Given the following observations, test the significance of the treatment differences (main effects) and interaction:

Variable I

		A	B	C
		4	3	2
		6	2	3
	a	8	2	2
		6	1	2
		3	4	4
		2	3	6
		3	4	8
Variable II	b	3	2	9
		2	3	4
		4	2	5
		3	9	2
		2	6	4
	c	3	5	3
		2	8	2
		2	4	2

Answers

1. The necessary totals can be obtained as indicated:

Variable I

		A		B		C		
		X	X^2	X	X^2	X	X^2	T
	a	9	81	8	64	9	81	26
Variable II	b	3	9	5	25	6	36	14
	c	6	36	6	36	4	16	16
	T =	18	126	19	125	19	133	56

$$\Sigma\Sigma X_{ij}^2 = 384$$

Total sum-of-squares:

$$\Sigma x_t^2 = \Sigma\Sigma X_{ij}^2 - \frac{T_{++}^2}{rc} = 384 - \frac{(56)^2}{9} = 384 - \frac{3136}{9}$$

$$= 384.00 - 348.44$$

$$= 35.56$$

Sum-of-squares for rows:

$$\Sigma x_{b_r}^2 = \Sigma \frac{T_{+j}^2}{n_j} - \frac{T_{++}^2}{rc} = \frac{(26)^2 + (14)^2 + (16)^2}{3} - 348.44$$

$$= \frac{1128}{3} - 348.44 = 376.00 - 348.44$$

$$= 27.56$$

Sum-of-squares for columns:

$$\Sigma x_{b_c}^2 = \Sigma \frac{T_{i+}^2}{n_i} - \frac{T_{++}^2}{rc} = \frac{(18)^2 + (19)^2 + (19)^2}{3} - 348.44$$

$$= \frac{1046}{3} - 348.44 = 348.67 - 348.44$$

$$= .23$$

Residual sum-of-squares:

$$\Sigma x_{res}^2 = \Sigma x_t^2 - (\Sigma x_{b_r}^2 + \Sigma x_{b_c}^2)$$
$$= 35.56 - (27.56 + .23) = 35.56 - 27.79$$
$$= 7.77$$

The sum-of-squares terms have the following degrees of freedom:
Total sum-of-squares: $rc - 1$, or $9 - 1 = 8$
Row means sum-of-squares: $r - 1$, or $3 - 1 = 2$
Column means sum-of-squares: $c - 1$, or $3 - 1 = 2$
Residual sum-of-squares: $(rc - 1) - (r - 1) - (c - 1)$, or $8 - 2 - 2 = 4$.

Summarizing the data in the conventional summary table below we obtain:

Summary Table

Source	Sum-of-squares	df	Mean square
Row means	27.56	2	13.78
Column means	.23	2	.12
Residual	7.77	4	1.94
Total	35.56	8	

The test for significance of differences among rows is:

$$F = \frac{MS_r}{MS_{res}}$$
$$= \frac{13.78}{1.94}$$
$$= 7.1$$

$P(F = 7.1)$ is smaller than 5 per cent for 2df and 4df, and we conclude that the differences between variable II (rows) categories is significant.

The test for significance of differences among columns is:

$$F = \frac{MS_c}{MS_{res}}$$

$$= \frac{.11}{1.94}$$

$$= .06$$

$P(F = .06)$ is greater than 5 per cent for 2df and 4df. Thus, we conclude that the variable I categories are not significantly different from each other.

2. The necessary totals are first computed:

Variable I

		A		B		C		
		X	X²	X	X²	X	X²	T
		4	16	3	9	2	4	
		6	36	2	4	3	9	
	a	8	64	2	4	2	4	
		6	36	1	1	2	4	
		3	9	4	16	4	16	
	T =	27	161	12	34	13	37	52
		2	4	3	9	6	36	
		3	9	4	16	8	64	
Variable II	b	3	9	2	4	9	81	
		2	4	3	9	4	16	
		4	16	2	4	5	25	
	T =	14	42	14	42	32	222	60
		3	9	9	81	2	4	
		2	4	6	36	4	16	
	c	3	9	5	25	3	9	
		2	4	8	64	2	4	
		2	4	4	16	2	4	
	T =	12	30	32	222	13	37	57
	T =	53		58		58		169

$$\Sigma\Sigma X_{ij}{}^2 = 827$$

Total sum-of-squares:

$$\Sigma x_t^2 = \Sigma\Sigma X_{ij}^2 - \frac{T_{++}^2}{rcn} = 827 - \frac{(169)^2}{45} = 827 - 634.69$$

$$= 192.31$$

Between-categories sum-of-squares:

$$\Sigma x_b^2 = \Sigma \frac{T_{ij}^2}{n_i} - \frac{T_{++}^2}{rcn} = \frac{(27)^2}{5} + \frac{(14)^2}{5} + \frac{(12)^2}{5} + \frac{(12)^2}{5} + \frac{(14)^2}{5} + \frac{(32)^2}{5}$$

$$+ \frac{(13)^2}{5} + \frac{(32)^2}{5} + \frac{(13)^2}{5} - 634.69$$

$$= \frac{3795}{5} - 634.69 = 759.00 - 634.69$$

$$= 124.31$$

Within-categories sum-of-squares:

$$\Sigma x_w^2 = \Sigma x_t^2 - \Sigma x_b^2$$

$$= 192.31 - 124.31$$

$$= 68.00$$

Column means sum-of-squares:

$$\Sigma x_{b_c}^2 = \Sigma \frac{T_{i+}^2}{rn} - \frac{T_{++}^2}{rcn} = \frac{(53)^2 + (58)^2 + (58)^2}{15} - 634.69$$

$$= \frac{9537}{15} - 634.69 = 635.8 - 634.69$$

$$= 1.11$$

Row means sum-of-squares:

$$\Sigma x_{b_r}^2 = \Sigma \frac{T_{+j}^2}{cn} - \frac{T_{++}^2}{rcn} = \frac{(52)^2 + (60)^2 + (57)^2}{15} - 634.69$$

$$= \frac{9553}{15} - 634.69 = 636.87 - 634.69$$

$$= 2.18$$

Interaction sum-of-squares:

$$\Sigma x_{int}^2 = \Sigma x_b^2 - (\Sigma x_{b_r}^2 + \Sigma x_{b_c}^2)$$
$$= 124.31 - (2.18 + 1.11)$$
$$= 121.02$$

Summary Table

Source	Sum-of-squares	df	Mean square
Rows	2.18	2	1.09
Columns	1.11	2	.56
Interaction	121.02	4	30.26
Subtotal (means)	124.31	8	15.54
Within	68.00	36	1.89
Total	192.31	44	

The significance of rows differences is:

$$F = \frac{MS_r}{MS_w}$$
$$= \frac{1.09}{1.89}$$
$$= .58$$

$P(F = .58)$ is greater than 5 per cent for 2df and 36df. Thus we conclude that there are no significant differences between variable II treatment effects.

The significance of columns differences is:

$$F = \frac{MS_c}{MS_w}$$
$$= \frac{.56}{1.89}$$
$$= .30$$

$P(F = .30)$ is greater than 5 per cent for 2df and 36df. We conclude that there are no significant differences between the variable I treatment effects.

The significance of interaction is:

$$F = \frac{MS_{int}}{MS_w}$$
$$= \frac{30.26}{1.89}$$
$$= 16.01$$

$P(F = 16.01)$ is less than 5 per cent for 4df and 36df. We conclude that the differences between cell means are significant at the α level and that the cell means are probably not normally distributed.

XI

Correlation

In previous chapters we examined methods used in testing hypotheses about *univariate* populations of events. That is, the observed events were obtained from distributions characterized by values of *one* variable.

The use of samples to estimate parametric values depends on the parametric assumptions, one of which, you will recall, stipulates that each observation be *independently* obtained from the population. By *independently* we mean that the observations in the population are not *correlated*.

In *correlated* populations, the selection of an observation of a variable X also determines the observation in the associated variable Y. For instance, let us suppose we wish to study the relationship between the height and weight of n individuals. We would begin by selecting, at random, a particular individual from the population and noting his height. This observation was obtained *at random*; however, his weight cannot be a random observation since selecting his height determines the observation of weight. Thus, *correlation* is a kind of bias.

Whenever paired observations are examined, selection of the X member of the pair determines the value of the Y member of the pair.

The correlation between two variables is the extent to which their values *vary together systematically*. It cannot be overemphasized that correlation is not the same thing as causality, though this is the common interpretation.

It cannot be denied that correlation between two variables is often used as an index of causality, or that causality is often implied by the correlation between sets of measures or observations. However, while correlation may in fact *reflect* causality, there is no statistical procedure which can be used to *establish* causality. Consequently, even if one variable did cause corresponding observations in another variable, we cannot know this from correlation analysis.

165

Suppose that you have a friend who, every day, at three o'clock, drinks a cup of tea. He has done this without fail for the last twenty-five years. On a particular day you predict to a mutual acquaintance that, since there is no obvious factor preventing your tea-drinking friend from pursuing his practice, and since it is now three o'clock, he will drink his usual cup of tea. The fellow then proceeds to do just that. Did your observation, or your prediction of the observation, cause your friend to drink the tea? Of course not.

There are many other examples of correlation between sets of events. However, the implications of this association must be obtained from the nature of the events and from the nature of the correlation analysis *per se*.

In order to clarify the distinction between correlation and causality further, we shall develop the idea that correlation analysis is a special case of the previously encountered analysis of variance design for the comparison of two groups, one-variable classification. In essence, the correlation coefficient, r_{XY}, is a special case of F.

Let us suppose that we have obtained two large samples: the first we obtained from population X and the second from population Y. When plotted, each of these two samples appears to be approximately normal in shape. Sample 1 has a mean $\bar{X}$ and a standard deviation S_X; sample 2 has a mean $\bar{Y}$ and a standard deviation S_Y. You will recall from previous discussions that

$$x = X - \bar{X}$$

and it therefore follows that

$$y = Y - \bar{Y}$$

Furthermore, we stipulate that, if the two distributions are identical, then both the X and the Y distributions are *normal*; that is, the Z values in the X and Y distributions have the same meaning $(Z_X = Z_Y)$: proportions of the area between corresponding Z values in the X and Y distributions are the same.

When two distributions are identical, then $X = Y$ and $x = y$. The product of the two deviations x and y is xy, and the sum of the products of the deviations, Σxy (called the *product sum,*) divided by df, must be an estimate of population variance. For this reason: if $X = Y$, and $x = y$, we may substitute X for Y. This substitution transforms the product sum from Σxy to Σxx. And, Σxx is, of course, Σx^2. Σx^2 divided by df, i.e. $\dfrac{\Sigma x^2}{\text{df}}$, is an unbiased estimate of σ^2.

The product sum, Σxy, when divided by df, gives the *sample covariance, c:*

$$c = \frac{\Sigma xy}{n - 1}$$

The sample covariance in correlated samples is analogous to the variance in independent samples.

The sample covariance divided by the product of the two standard deviations, S_X and S_Y, gives the *product-moment correlation coefficient, r_{XY}*:

$$r_{XY} = \frac{\dfrac{\Sigma xy}{n-1}}{S_X S_Y}$$

If the populations correlated are identical in distribution, and the samples obtained from these populations are likewise very similar in distribution, then $X = Y$, $x = y$, and we may substitute X for Y and x for y:

$$r_{XY} = \frac{\dfrac{\Sigma xx}{n-1}}{S_X S_X}$$

$$= \frac{\dfrac{\Sigma x^2}{n-1}}{S_X^2}$$

and since $\dfrac{\Sigma x^2}{n-1} = S_X^2$, we obtain

$$r_{XY} = \frac{S_X^2}{S_X^2}$$

$$= F$$

It can be seen from the substitution of terms above that the product-moment correlation coefficient, r_{XY}, is really a special case of the sampling distribution of the ratio of two estimates of population variance which we know to be distributed as F. The extent to which xy differs from x^2 is dependent on the differences between the distributions of the X and Y observations.

The Product-moment Correlation Coefficient: r_{XY}

In correlation analysis, we observe an event for each of two related variables describing a particular individual in the population or sample. The use of the product-moment correlation coefficient as an index of association or dependence is based on the assumption that the observations in each distribution, X and Y, were obtained from normally distributed populations. This means, among other things, that the variables X and Y are continuous rather than discrete. It is important to keep this in mind. For example, there is undoubtedly some degree of correlation between subject sex and weight. However,

r_{XY} cannot be used to determine that correlation, since sex is a discrete (dichotomous) variable.

Empirically speaking, it is relatively unusual to find normally distributed populations that are bivariate. However, if the assumption of normality is reasonably well satisfied, r_{XY} becomes a test of independence of the two populations.

The relationship between distributions of observations of two variables may be described in terms of the *regression line* (or *prediction line*), which may be used to determine values of one variable from values of the other. Regression lines, based upon the correlation between the X and Y variables, are lines that, in a graphic representation of the X and Y variables, describe the relationship between points on the X variable predicted from Y and points on the Y variable predicted from X.

Independence, as applied in correlation, is important because it indicates that correlation is of primary value in terms of its contribution to prediction of values of Y from the X distribution and values of X from the Y distribution.

It should be pointed out here that the coefficient of correlation is of no practical value to the average investigator. We may set up criteria to test hypothetical differences between obtained values of r_{XY} and the value of the population parameter, ρ (Greek rho), but this will not necessarily tell us much about the *degree* of dependence of the bivariates. Correlational procedures, as previously indicated, serve a function in prediction. This will become more apparent when we examine regression later on.

The extent to which two variables are correlated is the extent to which we may reliably predict values of one variable from observations of the other. For instance, that height and weight are correlated variables is of consequence only if we wish to predict height from weight or weight from height. Again, this type of prediction is easily misunderstood as indicating causality.

The Correlation Coefficient r_{XY} as an Index of Linear Association

In order to determine the relationship between bivariate distributions, a *scatter diagram* may be employed. On this scatter diagram, each XY point is plotted. (You will recall that, in previous graphic methods, the abscissa of the graph represented categories of the variable, X, and the ordinate of the graph typically represented the frequency with which each category was represented in the distribution.) The scatter diagram is a kind of *multi-parametric graph:* the abscissa represents the X variable and the ordinate represents the Y variable. Thus both abscissa and ordinate represent a scale of variable categories. The XY points are represented as points at the intersection of the graph co-ordinates. The position on the graph of each XY point makes it possible to determine its X value and its Y value.

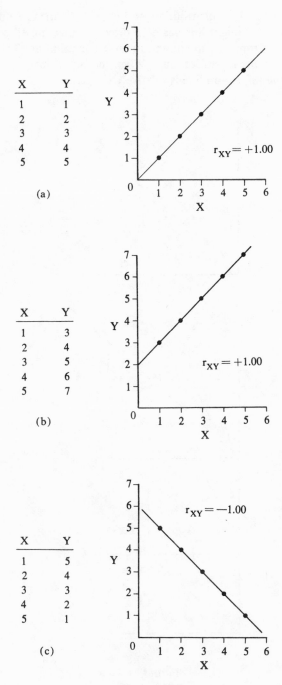

X	Y
1	1
2	2
3	3
4	4
5	5

(a)

$r_{XY} = +1.00$

X	Y
1	3
2	4
3	5
4	6
5	7

(b)

$r_{XY} = +1.00$

X	Y
1	5
2	4
3	3
4	2
5	1

(c)

$r_{XY} = -1.00$

FIGURE 11.1

In the case of the three distributions in Figure 11.1, the XY points lie in a straight line, and a straight line has been drawn connecting all points. Note that the distance from each point and the line is minimum. This line is the line of best fit, or the regression line. When the XY points lie on a straight line, drawing the regression line is relatively simple.

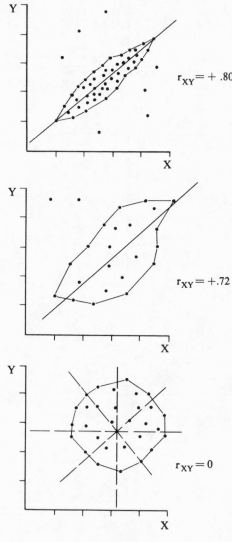

FIGURE 11.2

In the first distribution, (a), the Y values increase as the X values increase. The resulting slope of the regression line is said to indicate positive correlation. In the second distribution, r_{XY} is the same, but the intercept is different. In the third distribution, (c), the slope is perpendicular to the slope of the regression line in the first distribution. This results in negative correlation.

Note that the sign of the correlation coefficient, plus or minus, simply indicates the sign of the slope of the regression line. The coefficient itself may have any value ranging between 0.00 and 1.00. Thus it is possible to obtain values between -1.00 and $+1.00$. The departure of the value of the coefficient from its maximum value, i.e. $+1.00$, is related to the departure of the XY points in the scatter diagram from a straight line.

Thus, the correlation coefficient indicates the extent to which the relationship between the bivariates is *linear*. A correlation coefficient of 1.00 indicates perfect *linearity*, while a correlation coefficient of 0.00 indicates perfect *non-linearity*, which is perfect *curvilinearity*.

In most cases, empirically obtained bivariates seldom result in XY points that lie on a perfect straight line in the scatter diagram. There is a rapid though not very reliable way of estimating the degree of correlation from the scatter diagram in such a case. The regression line may be "fitted" in the following way: We draw a line connecting the XY points on the perimeter of the densest portion of the scatter diagram, so as to form an envelope surrounding the XY points that constitute this densest portion. This envelope is then bisected longitudinally. The line that bisects the envelope is the visually fitted regression line. Examples of this type of correlation estimation are shown in Figure 11.2. Computed values of r_{XY} for these distributions are shown as well. Note that when $r_{XY} = 0.00$, it is virtually impossible to determine how one should bisect the structure "longitudinally."

The line of best fit determined visually is almost always drawn so that the portion of the scatter diagram that contains the greatest concentration of XY points is longitudinally bisected. By bisecting the densest portion longitudinally, we make the distance from the line of each XY point (i.e. its deviation from the regression line) the minimum possible. There is a mathematical method for establishing a regression line so that it will have this property; it is called the *method of least squares*. We will discuss it later in this chapter.

Computation of r_{XY} from Original Observations

The product-moment correlation coefficient can be computed directly from the raw data. We will use the data in Table 11.1 to illustrate the computational procedures.

X	x	x^2	X^2	Y	y	y^2	Y^2	xy	XY
40	+5	25	1600	44	+12	144	1936	60	1760
39	+4	16	1521	38	+6	36	1444	24	1482
38	+3	9	1444	36	+4	16	1296	12	1368
38	+3	9	1444	33	+1	1	1089	3	1254
36	+1	1	1296	32	0	0	1024	0	1152
35	0	0	1225	29	−3	9	841	0	1015
34	−1	1	1156	28	−4	16	784	4	952
33	−2	4	1089	27	−5	25	729	10	891
29	−6	36	841	27	−5	25	729	30	783
28	−7	49	784	26	−6	36	676	42	728
350	0	150	12,400	320	0	308	10,548	185	11,385

$$\Sigma X = 350 \qquad\qquad\qquad \Sigma Y = 320$$

$$\bar{X} = \frac{\Sigma X}{n} = \frac{350}{10} = 35.0 \qquad\qquad \bar{Y} = \frac{\Sigma Y}{n} = \frac{320}{10} = 32$$

$$S_X^2 = \frac{\Sigma x^2}{n} = \frac{150}{10} = 15 \qquad\qquad S_Y^2 = \frac{\Sigma y^2}{n} = \frac{308}{10} = 30.8$$

$$S_X = \sqrt{15} = 3.87 \qquad\qquad S_Y = \sqrt{30.8} = 5.55$$

TABLE 11.1

The obtained values may be substituted in the formula:

$$r_{XY} = \frac{\dfrac{\Sigma xy}{n-1}}{S_X S_Y}$$

$$= \frac{\dfrac{185}{9}}{3.87(5.55)}$$

$$= \frac{20.56}{21.48}$$

$$= .957 \quad \text{or} \quad .96$$

The calculator formula may also be used:

$$r_{XY} = \frac{\left[\Sigma XY - \dfrac{(\Sigma X)(\Sigma Y)}{n}\right] \Big/ n - 1}{\sqrt{\left(\dfrac{\Sigma X^2 - \dfrac{(\Sigma X)^2}{n}}{n}\right)\left(\dfrac{\Sigma Y^2 - \dfrac{(\Sigma Y)^2}{n}}{n}\right)}}$$

$$= \frac{\dfrac{11385 - 11200}{9}}{\sqrt{\left(\dfrac{150}{10}\right)\left(\dfrac{308}{10}\right)}}$$

$$= \frac{20.56}{21.50}$$

$$= .956 \quad \text{or} \quad .96$$

The data in the distribution may be plotted in a scatter diagram and a line of best fit may be drawn to approximate the regression line. The scatter diagram and visually estimated regression line are shown in Figure 11.3.

It is profitable to return for a moment to the calculator formula for r_{XY}. It should be noted that this formula is simply the substitution of the alternate

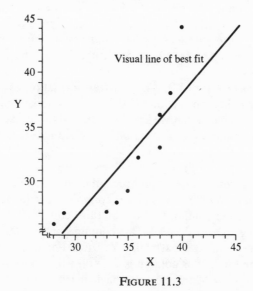

FIGURE 11.3

method for computing Σx^2. The denominator of this formula would be, in the context of our discussion in the beginning of this chapter:

$$\frac{\Sigma XX - \dfrac{(\Sigma X)(\Sigma X)}{n}}{n-1}$$

or

$$\frac{\Sigma X^2 - \dfrac{(\Sigma X)^2}{n}}{n-1}$$

which is an unbiased estimate of σ^2.

Computation of r_{XY} from a Frequency Table

It is often desirable to determine the correlation between samples constituted of relatively large numbers of paired observations. To do this we group observations and compute r_{XY} from grouped data. Although the data in Table 11.1 do not constitute a large number of observations, we will use them in the illustrative example below for the sake of simplicity.

The data in the frequency table, Table 11.2, have been grouped with $i = 2$ class intervals. There are ten intervals. Ordinarily, a sample of so small an n would not be grouped; grouped data invariably reduce the effectiveness of a test because of loss of information and subsequent distortion. However, with a reasonably small n, the computational procedures are simpler to follow. A detailed explanation of each column follows. The hash marks (slashes) represent the XY points in the appropriate class interval.

First column, upper right (f). This column contains the frequency in each of the YX rows.

Second column, upper right (Y'). Y' is a transformation of the Y class-interval midpoint by the subtraction of the lowest interval midpoint and division by i. (You will recall that we first encountered this transformation in connection with computation of the variance and standard deviation from grouped data.)

Third column, upper right (fY'). The values in this column are obtained by multiplying the values in the first and second columns: $(f)(Y')$.

Fourth column, upper right (fY'^2). In this column, the values are obtained by multiplying the second column (Y') by the third column (fY').

Fifth column, upper right ($\Sigma X'._{Y'}$). This column contains the sum of the X' values that correspond to a particular Y' row. For instance, the value in the first cell at the top of this column is 7. You will note that, in the row that corresponds to $Y' = 9$, there is one hash mark in the $Y = 44$ to 45

and X = 40 to 41 portion of the scatter diagram. The X' value that corresponds to the noted entry (Y' = 9) is 7.

Examine the row in the scatter diagram Y = 32 to 33. There is one hash
mark in the column X = 36 to 37 and one hash mark in the column X = 38
to 39. The Y' value that corresponds to the Y = 32 to 33 row is 3. The
first hash mark in the Y = 32 to 33 row and X = 36 to 37 column has an X'
value of 5; the second hash mark in the Y = 32 to 33 row and the X = 38
to 39 column has an X' value of 6. The sum of these two X' values is 5 + 6
= 11. The value 11 is entered in the $\Sigma X'._{Y'}$ column, in the row corresponding
to Y = 32 to 33.

Sixth column, upper right ($Y'\Sigma X'._{Y'}$). The values in this column result
from multiplying column five ($\Sigma X'._{Y'}$) by column two (Y').

For additional illustration, let us examine the row corresponding to Y
= 28 to 29. Note that in the scatter diagram there are two hash marks
corresponding to X = 34 to 35 for that Y row. In the first column, f, the
frequency in the row is entered. The frequency is 2. In the second column,
the Y' value is 1. Thus, in the third column, fY', the value 2 is entered. The

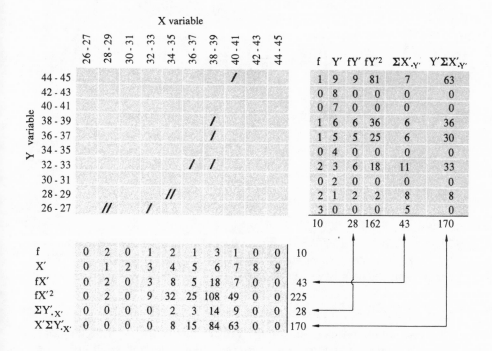

TABLE 11.2

fourth column is the product of column two and three, respectively: $Y' = 1$, $fY' = 2$, and $(1)(2) = 2$. In the fifth column ($\Sigma X'._{Y'}$), the value 8 is obtained in the following way: The two hash marks in scatter diagram each have an X' value of 4. Thus, the sum of the X' values is 8. Finally, the value in the sixth column is obtained by multiplying the values in column two and five, respectively: 1 and 8.

Below the scatter diagram there are six rows whose computations are analogous to those of the six columns just described. These rows are composed of appropriate Y values, while the columns refer to X values. Since the computations are the same, we will not repeat them.

The product-moment correlation coefficient, r_{XY}, can be computed from the frequency table:

$$r_{XY} = \frac{\Sigma X'Y' - \dfrac{(\Sigma fX')(\Sigma fY')}{n}}{\sqrt{\left(\Sigma fX'^2 - \dfrac{(\Sigma fX')^2}{n}\right)\left(\Sigma fY'^2 - \dfrac{(\Sigma fY')^2}{n}\right)}}$$

$$= \frac{170 - \dfrac{(43)(28)}{10}}{\sqrt{\left(225 - \dfrac{(43)^2}{10}\right)\left(162 - \dfrac{(28)^2}{10}\right)}}$$

$$= \frac{170 - 120.4}{\sqrt{(225 - 184.9)(162 - 78.4)}}$$

$$= \frac{49.60}{\sqrt{3352.36}}$$

$$= .85$$

The obtained coefficient, .85, is somewhat smaller than the coefficient obtained by computation from raw data. This difference is due to the small size of the sample and the data loss in grouping.

Computation of r_{XY} from Standard Scores

This computational method, though rather tedious and of little empirical value, nevertheless is interesting because it illustrates the relationship between the populations in bivariate distributions.

Table 11.3 presents the data necessary for the computation of r_{XY} from standard scores (z), and it is based on the distributions in Table 11.1.

X	Y	x	y	z_X	z_Y	z_Xz_Y
40	44	+5	+12	+1.29	+2.16	+2.79
39	38	+4	+6	+1.03	+1.08	+1.11
38	36	+3	+4	+0.78	+0.72	+0.56
38	33	+3	+1	+0.78	+0.18	+0.14
36	32	+1	0	+0.26	0.00	0.00
35	29	0	−3	0.00	−0.54	0.00
34	28	−1	−4	−0.26	−0.72	+0.19
33	27	−2	−5	−0.52	−0.90	+0.47
29	27	−6	−5	−1.55	−0.90	+1.40
28	26	−7	−6	−1.81	−1.08	+1.95
$\Sigma = 350$	$\Sigma = 320$					$\Sigma = +8.61$

TABLE 11.3

The coefficient is then computed as follows:

$$r_{XY} = \frac{\Sigma z_X z_Y}{n}$$

and, substituting the values obtained in Table 11.3,

$$r_{XY} = \frac{8.61}{10}$$

$$= .86$$

In the above definition, the r_{XY} coefficient is the mean of the z products.

Linear Regression

Scientific investigators study populations in order to understand relationships between observed events. They often wish to understand the rules that describe the probability of observing certain kinds of events. These rules are often guides for prediction of the probability of the occurrence of events in a universe of alternatives. Some of them have already been examined.

In bivariate distributions, observations are not independent: correlation indicates the extent to which observations in the X distribution may be used to determine the probable value of corresponding observations in the Y distribution. This kind of bias permits a different kind of prediction.

The prediction made possible by correlation is based on *regression*. The regression, from an X value to a Y value, is a prediction: for each X value there is a corresponding Y value. When correlation is perfectly linear ($+1.00$ or -1.00), the *true* Y value, predicted from an X value, and the *predicted* Y value, $\tilde{Y}$, will be the same, i.e. $Y - \tilde{Y} = 0$.

Correlation between two variables is, however, seldom perfectly linear. There are two ways in which observations may fluctuate to result in a coefficient whose value is not 1.00. First, the relationship between the variables is not linear: one distribution is normal and the other is markedly skewed. Second, the corresponding X and Y values do not fall near the regression line and variance in the distributions is greater than might be normally expected. Examples of linear and non-linear relationships between hypothetical X and Y distributions are presented in Figure 11.4.

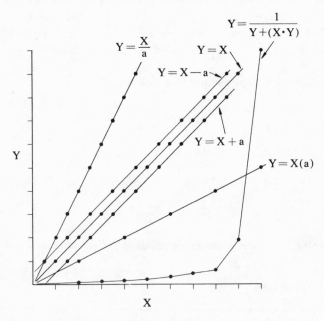

FIGURE 11.4

Usually, in empirically obtained distributions, the XY points do not fall in the same plane and consequently they cannot be connected by one single straight line. This is why it is necessary to determine the regression line by the *"least squares method."* This method minimizes the sum of the squared deviations of the XY points from the regression line.

The Line of Best Fit: Method of Least Squares

In Figure 11.3, the data in Table 11.1 were used to draw a "line of best fit," estimating the regression between the X and Y variables. This approximation by visual inspection attempts to take into account the fact that the line must somehow be such that the distance of each XY point from the line is the smallest possible. This procedure results in an estimate of the appropriate *slope* and *intercept* of a line which meets this requirement.

The *slope* of the regression line is the angle it forms with the base of the scatter diagram, and the *intercept* is the point on the ordinate where it originates. Figure 11.4 shows three regression lines with the same slope but different intercepts. These three lines are identified by the mathematical relation between the X and Y distribution values and are, respectively, $Y = X - a$, $Y = X$, and $Y = X + a$, where a is a constant value.

The line of best fit is that line which, based on values of X, produces the best estimate of corresponding values of Y. The difference between the estimated Y value ($\tilde{Y}$) and the actual Y value (Y) can be defined as $\tilde{Y} - Y$. The line of best fit is a line that makes this difference the smallest obtainable.

To determine the line of best fit, let us examine the properties of a line— any line—not parallel to the base of the scatter diagram. Such a line may be defined as

$$Y = a + bX$$

Graphically represented, such a line has its Y intercept at a, and has a slope b, for any values of the constants a and b. When the values of a and b are known, the line is completely determined and can be used to establish the $\tilde{Y}$ from each value of X. Thus,

$$\tilde{Y} = a + bX$$

To define "best fit," we say that the difference $\tilde{Y} - Y$ is the error of prediction resulting from the fact that we have used X values to determine Y values, rather than using actual Y values, and that the sum of the predictive errors for the "line of best fit" is the smallest obtainable.

For practical reasons, the "line of best fit" is defined in terms of the *squared deviations* of the XY points from the regression line. (The "line of best fit" is an *average*, and the sum of the deviations of the XY points about this line must, therefore, be zero.)

The method of least squares consists in determining the constants a and b in the linear equation describing the relationship between the distributions in question.

The constant b, known as the *regression coefficient*, is defined as

$$b = \frac{\Sigma XY - n\bar{X}\bar{Y}}{\Sigma x^2 - n\bar{X}^2}$$

$$= \frac{\Sigma xy}{\Sigma x^2}$$

The intercept, a, is defined as

$$a = \bar{Y} - b\bar{X}$$

Thus the regression line is defined as

$$\tilde{Y} = (\bar{Y} - b\bar{X}) + bX$$

or

$$\tilde{Y} = \left[\bar{Y} - \left(\frac{\Sigma xy}{\Sigma x^2}\right)\bar{X}\right] + \left(\frac{\Sigma xy}{\Sigma x^2}\right)(X)$$

The line of best fit can be determined for the data in Table 11.1 by substituting the appropriate values:

$$b = \frac{\Sigma xy}{\Sigma x^2}$$

$$= \frac{185}{150}$$

$$= 1.23$$

and

$$a = \bar{Y} - b\bar{X}$$

$$= 32 - (1.23)(35)$$

$$= 32 - 43.05$$

$$= -11.05$$

and thus,

$$\tilde{Y} = (\bar{Y} - b\bar{X}) + bX$$

$$= (-11.05) + 1.23(X)$$

Only two points are necessary to determine a line. For illustrative purposes, we choose two points in the X distribution, X = 40 and X = 28.

When X = 40,

$$\tilde{Y} = (-11.05) + 1.23(40)$$
$$= (-11.05) + (49.20)$$
$$= 38.15$$

and when X = 28,

$$\tilde{Y} = (-11.05) + 1.23(28)$$
$$= (-11.05) + 34.44$$
$$= 23.39$$

Figure 11.5 presents a comparison of the visual fit regression line and the regression line obtained by the least squares method.

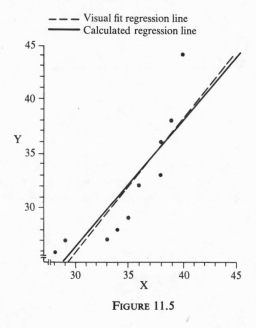

FIGURE 11.5

In Figure 11.6, three predictive lines are drawn. The first is $\tilde{Y}$ based on the actual value of Y, the second is $\tilde{Y}$ based on visual fit regression line, and the third is $\tilde{Y}$ based on the least squares regression line.

The X value of 34, from which Y values are predicted, has an actual Y value of 28. The predicted Y value, using the visually estimated regression line, is approximately 30.6, while the Y value estimated from the least squares regression line is 31.0. It is easy to see that the closer the XY points are to the regression line, the closer will be the predicted Y value based upon any value of X.

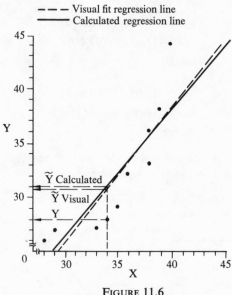

FIGURE 11.6

The Regression of Y on X

So far our discussion has concerned the regression of X on Y. It is also possible to predict values of X from values of Y. If we can predict weight from height, why not height from weight? The regression of Y on X is simply this reverse process.

$$b_{XY} = \frac{\Sigma xy}{\Sigma x^2}$$

$$b_{YX} = \frac{\Sigma xy}{\Sigma y^2}$$

Substituting the data in Table 11.1,

$$b_{YX} = \frac{185}{308}$$

$$= .60$$

and

$$a_{XY} = \overline{Y} - b_{XY}\overline{X}$$
$$a_{YX} = \overline{X} - b_{YX}\overline{Y}$$
$$= 35.0 - (.60)(32.0)$$
$$= 35.0 - 19.2$$
$$= 15.8$$

then

$$\tilde{Y} = (\bar{Y} - b_{XY}\bar{X}) + b_{XY}(X)$$

and

$$X = (\bar{X} - b_{YX}\bar{Y}) + b_{YX}(Y)$$
$$= 15.80 + (.60)(Y)$$

Two values of Y can be selected to determine the line. For illustrative purposes, the values are $Y = 44$ and $Y = 28$. When $Y = 44$,

$$X = 15.80 + (.60)(44)$$
$$= 15.80 + 26.40$$
$$= 42.20$$

and when $Y = 28$,

$$X = 15.80 + (.60)(28)$$
$$= 15.80 + 16.80$$
$$= 32.60$$

These two points determine the regression line for prediction of X values from Y values. Both regression lines are presented in Figure 11.7.

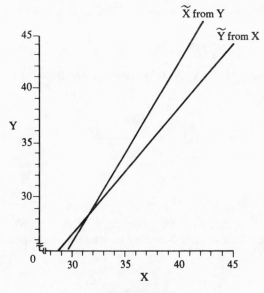

FIGURE 11.7

If the X and Y distributions are identical, then the points that determine the regression of X on Y and the points that determine the regression of Y on X should lie in the same plane, and only one straight line, and no other, will describe both. When the regression lines are maximally different, they will be at right angles. When the regression lines are at right angles, the distributions are said to be independent, or *orthogonal*, and r_{XY} approaches zero.

The Standard Error of Estimate

Our discussion of regression should lead to the conclusion that the regression line is a sort of average: an average line of prediction. The XY points in the distribution deviate from the regression line by some quantity. The variance of the XY observations around the regression line is denoted as S_{YX}^2. The square root of this term, S_{YX}, is called the *standard error of estimate*.

The standard error of estimate is the standard deviation of a distribution whose mean is the line determined by the predicted Y values. It is defined as

$$S_{YX} = \sqrt{\frac{\Sigma(Y - \tilde{Y})^2}{n - 2}}$$

When the correlation coefficient approaches 1.00, then S_{YX} approaches zero. And when r_{XY} approaches zero, S_{YX} approaches S_Y, the standard deviation of the Y distribution. The simplest computation of S_{YX} is given below.

$$S_{YX} = \sqrt{\frac{\Sigma y^2 - \frac{(\Sigma xy)^2}{\Sigma x^2}}{n - 2}}$$

The data in Table 11.1 may be substituted to illustrate the computation of S_{YX}.

$$S_{YX} = \sqrt{\frac{308 - \frac{(185)^2}{150}}{8}}$$

$$= \sqrt{\frac{79.83}{8}}$$

$$= 3.16$$

Values in the X distribution predicted from Y also results in a standard error of estimate. This is determined by solving the corresponding formula:

$$S_{XY} = \sqrt{\frac{\Sigma x^2 - \frac{(\Sigma xy)^2}{\Sigma y^2}}{n - 2}}$$

Substituting the data in Table 11.1, we obtain

$$S_{XY} = \sqrt{\dfrac{150 - \dfrac{(185)^2}{308}}{8}}$$

$$= \sqrt{\dfrac{38.88}{8}}$$

$$= 2.21$$

One important consideration in the application of regression, in the context of the discussion of S_{YX} and S_{XY}, is the precise nature of the prediction made. We have used the terms $\tilde{Y}$ and X as though these were predicted *point* values in their respective distributions. Actually, the Y value predicted from X and the X value predicted from Y are not really points at all: each is the

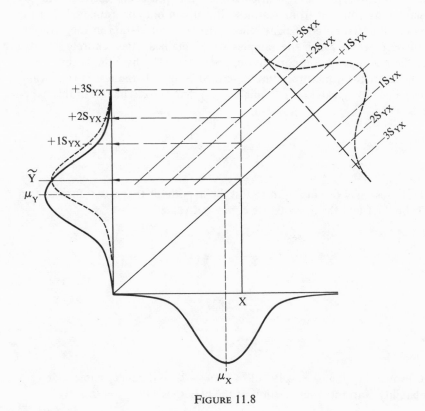

FIGURE 11.8

mean of a predicted distribution. For instance, given the specific weight of an individual, and the correlation between height and weight, it is not possible to predict his *precise* height; it is merely possible to predict the *average* height of individuals with that weight. Conversely, it is possible to predict the *average* weight of individuals with a given height, but not the *exact* weight of a *particular* individual.

Figure 11.8 illustrates the fact that the prediction of a Y value from a specific X value is the mean of a predicted distribution.

The mean of the X distribution is μ_X and the mean of the Y distribution is μ_Y. A specific X value is chosen. Based on this X value, a specific point is determined on the base of the Y distribution: that point corresponds to $\tilde{Y}$, the predicted Y value. And $\tilde{Y}$ is the mean of the distribution predicted.

Testing the Hypothesis that $r = \rho = 0$

The correlation procedures discussed in the preceding sections employ observations obtained from samples. But it can only be estimated that the obtained correlation coefficient reflects the true correlation in the bivariate populations from which the samples were obtained. It is entirely possible that the true population correlation, ρ, is zero. The hypothesis that $\rho = 0$ can be tested with the procedure described below. If the test is significant, ρ is significantly greater $(+)$ or smaller $(-)$ than zero. The test of this hypothesis is

$$t = \frac{r_{XY}}{\sqrt{1 - r_{XY}^2}} \sqrt{n - 2}$$

This *t* test may be employed to test the hypothesis that $\rho = 0$ for the data in Table 11.1 and the appropriate computed values:

$$t = \frac{.83}{\sqrt{1 - .69}} (\sqrt{8})$$

$$= \frac{.83}{.56} (2.83)$$

$$= 4.19$$

The probability that $t = 4.19$, df $= 8$ $(n - 2)$, is smaller than 5 per cent probability warrants the rejection of the hypothesis that $\rho = 0$.

Problems

1. Given the following 20 pairs of observations:

X	Y
47	35
37	35
34	31
29	29
27	26
25	24
25	22
23	22
23	21
22	20
21	19
21	16
18	16
17	16
13	15
11	14
8	11
7	10
6	9
1	1

a. Plot the observations on a scatter diagram and determine the regression line by visual estimation.

b. Determine the correlation between the variables.

c. Employing the method of least squares, determine the regression of X on Y.

d. Test the hypothesis that $\rho = 0$.

Answers

1. a. The scatter diagram:

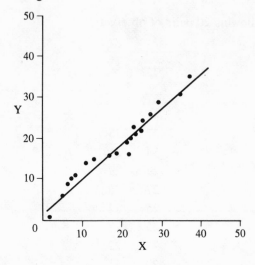

b. r_{XY}:

X	X²		Y	Y²		XY
47	2209		35	1225		1645
37	1369		35	1225		1295
34	1156		31	961		1054
29	841		29	841		841
27	729		26	676		702
25	625		24	576		600
25	625		22	484		550
23	529		22	484		506
23	529		21	441		483
22	484		20	400		440
21	441		19	361		399
21	441		16	256		336
18	324		16	256		288
17	289		16	256		272
13	169		15	225		195
11	121		14	196		154
8	64		11	121		88
7	49		10	100		70
6	36		9	81		54
1	1		1	1		1
$\Sigma =$ 415	11031		392	9166		9973

$$\bar{X} = \frac{415}{20} = 20.75 \qquad\qquad \bar{Y} = \frac{392}{20} = 19.60$$

$$(\Sigma X)^2 = (415)^2 \qquad\qquad (\Sigma Y)^2 = (392)^2$$
$$ = 172225 \qquad\qquad = 153664$$

$$\Sigma x^2 = \Sigma X^2 - \frac{(\Sigma X)^2}{n} \qquad\qquad \Sigma y^2 = \Sigma Y^2 - \frac{(\Sigma Y)^2}{n}$$

$$ = 11031 - \frac{172225}{20} \qquad\qquad = 9166 - \frac{153664}{20}$$

$$ = 11031 - 8611.25 \qquad\qquad = 9166 - 7683.20$$

$$ = 2419.75 \qquad\qquad = 1482.80$$

$$r_{XY} = \frac{\Sigma XY - \dfrac{(\Sigma X)(\Sigma Y)}{n}}{(\sqrt{\Sigma x^2})(\sqrt{\Sigma y^2})}$$

$$= \frac{9973 - \dfrac{(415)(392)}{20}}{(\sqrt{2419.75})(\sqrt{1482.8})}$$

$$= \frac{1839}{(49.19)(38.51)}$$

$$= .97$$

c. The regression of X on Y:

$$\tilde{Y} = \left[\bar{Y} - \frac{\Sigma xy}{\Sigma x^2}(\bar{X}) \right] + \frac{\Sigma xy}{\Sigma x^2}(X)$$

$$= \left[19.60 - \frac{1839.00}{2419.75}(20.75) \right] + \frac{1839.00}{2419.75}(X)$$

$$= 19.60 - .76(20.75) + .76(X)$$

$$= 3.83 + .76(X)$$

Choosing two values of X, 29 and 13, we obtain

$$\tilde{Y} = 3.83 + .76(29)$$
$$= 3.83 + 22.04$$
$$= 25.87$$

and

$$\tilde{Y} = 3.83 + .76(13)$$
$$= 3.83 + 9.88$$
$$= 13.71$$

These points may be plotted and the regression line determined:

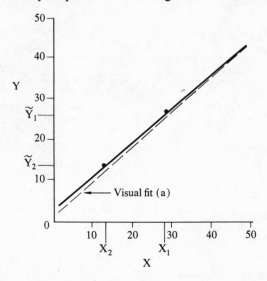

d. Testing the hypothesis that $\rho = 0$:

$$t = \frac{r_{XY}}{\sqrt{1 - r_{XY}^2}} (\sqrt{n - 2})$$

$$= \frac{.97}{\sqrt{1 - (.97)^2}} (\sqrt{18}) = \frac{.97}{.245} (4.24) = 16.79$$

The probability of $t = 16.78$, df $= 18$, is less than 5 per cent.

XII

Correlation II: Measures of Association

In our discussion of correlation, we said that the product-moment correlation coefficient, r_{XY}, was the appropriate measure of association between two variables that have the following properties: the observations are normally distributed, and they constitute a continuous variable. Under certain circumstances, one—or both—of these conditions is not met by empirically obtained observations.

When the criteria for the applicability of r_{XY} are *not* met, other measures of association between distributions may sometimes be employed.

The Biserial Correlation Coefficient: r_b

The biserial correlation coefficient may be used to determine the degree of association between variables that have the following property: observations are obtained from members of a bivariate population consisting of continuous and normally distributed observations in which one of the distributions is arbitrarily divided to form a dichotomous variable. That is, for instance, all observations above a certain value in the distribution are indicated as "passing" and all observations below that point are indicated as "failing"—in that they meet or fail to meet a certain criterion.

Thus, while both variables are continuous and normally distributed, one variable is now reduced to a dichotomy. Such a procedure might be employed in the comparison of the performance of given individuals on two tests, X and Y. For test X, actual scores are given. For test Y, individuals who score below a certain point fail, while those who score above a certain point pass. It might then be desirable to determine whether there is a relationship between performance on test X and performance on test Y.

191

To illustrate the computational procedure in determining r_b, we will use the distributions in Table 12.1. The X variable is continuous in distribution, and the Y variable is dichotomous.

X	Y
5	16
9	33
19	38
13	43*
27	45*
21	41*
35	40*
12	36
35	47*
41	43*
50	40*
48	45*
61	31
33	34
57	39
58	26
63	26
28	38
11	18
43	47*
$\Sigma X = 669$	$\Sigma Y = 726$

TABLE 12.1

$$r_b = \frac{\bar{X}_p - \bar{X}_T}{S_X}\left(\frac{p}{Y}\right)$$

where $\bar{X}_p$ is the mean of the distribution of X values that correspond to Y values scored as passing. In our example, $\bar{X}_p = 34.77$ and $\bar{X}_T$ is the mean of the entire X distribution, i.e. 33.45. The term p is the proportion of cases in the passing category, $p = .45$, Y is the ordinate of the normal distribution for a given proportion p of the area where $p + q = 1.00$, and $\frac{p}{Y}$ is a value obtained from Table A.10, p. 277. This value is 1.1369. The term S_X is, of

(*scored as passing)

course, the standard deviation of the X distribution, 18.28. Substituting the obtained data,

$$r_b = \frac{\bar{X}_p - \bar{X}_T}{S_X}\left(\frac{p}{Y}\right)$$

$$= \frac{34.77 - 33.45}{18.71}(1.1369)$$

$$= \frac{1.32}{18.71}(1.1369)$$

$$= .08$$

If the correlation between the distributions is relatively high, the difference between the means of the passing and failing scores should be approximately equal to the difference between the mean of the X distribution associated with passing and that associated with failing, and that difference should be maximum. However, if correlation is relatively low, the difference between these means should be approximately zero.

The Point Biserial Correlation Coefficient: r_{pb}

In biserial correlation, the dichotomous variable was originally a continuous variable. It is often desirable to determine the relationship between a continuous variable and a truly discrete variable which, by definition, cannot possibly be normally distributed. For instance, is there a relationship between test performance and sex of the test takers? Or, perhaps, is perceptual motor skill related to color blindness? In such cases, the point biserial correlation coefficient may be employed.

Computing r_{pb}

The point biserial correlation coefficient is computed as follows:

$$r_{pb} = \frac{\bar{X}_p - \bar{X}_T}{S_X}\left(\sqrt{\frac{p}{Y}}\right)$$

To illustrate the computation, let us assume that the same data used to compute r_b meets the requirements for this type of analysis. Substituting the obtained data,

$$r_{pb} = \frac{34.77 - 33.45}{18.71}(1.07)$$

$$= .08$$

The terms in this formula are the same as those in the formula used in the computation of r_b.

The Correlation Ratio: η

The product-moment correlation coefficient, r_{XY}, is a measure of the degree of linearity of the association between two variables, X and Y. Its quantitative aspects do not, however, reveal the extent of the departure of the relevant variables from linear association: a scatter diagram in which the points X and Y are distributed in random fashion is quite different from a scatter diagram in which the points show relatively little deviation from a line of best fit, which is curvilinear. Thus, while there is no mathematical assumption stipulating that the correlated data must be linearly related, the coefficient will reflect curvilinearity.

When data are known to be related in a non-linear manner, it is desirable to employ an index of association which is applicable to curvilinear *trends*. The appropriate index is the correlation ratio η (Greek eta).

In order to illustrate the computation of η and show its relationship to the product-moment correlation coefficient, r_{XY}, we shall use the data in Table 12.1. The data consist of 20 XY values. To reduce the number of computing categories—and for other reasons which we shall not develop here—the data are plotted as frequencies per class interval. It is possible, however, to compute both r_{XY} and η directly from the data. The frequency distribution by class intervals is plotted in Figure 12.1.

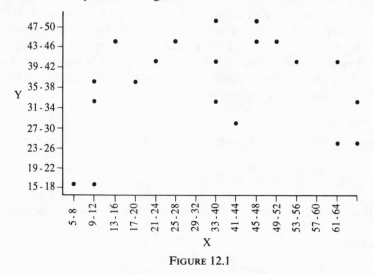

FIGURE 12.1

Note that in Figure 12.1 the line of best fit is clearly not a straight line. That is, a straight line *could* be drawn to fit the data so that there would be equal weight of XY points above and below that line, but such a line would not have least squares property. To have least squares property, a line to fit these data must be curved.

If we were not aware that curvilinearity affects the product-moment correlation coefficient, r_{XY}, we might compute such a coefficient. The computation of r_{XY} from Table 11.2, a correlation table (grouped data), is shown below.

$$r_{XY} = \frac{\Sigma X'Y' - \dfrac{(\Sigma X')(\Sigma Y')}{n}}{\sqrt{\Sigma X'^2 - \dfrac{(\Sigma X')^2}{n}}\sqrt{\Sigma Y'^2 - \dfrac{(\Sigma Y')^2}{n}}}$$

Substituting our data, we obtain

$$r_{XY} = \frac{706 - \dfrac{(139)(97)}{20}}{\sqrt{1342 - \dfrac{(139)^2}{20}}\sqrt{583 - \dfrac{(97)^2}{20}}}$$

$$= \frac{706 - \dfrac{13483}{20}}{\sqrt{1342 - \dfrac{19321}{20}}\sqrt{583 - \dfrac{9409}{20}}}$$

$$= \frac{31.85}{(21.82)(21.69)} = \frac{31.85}{473.28}$$

$$= .155 \quad \text{or} \quad .16$$

(You may wish to test the hypothesis that $r_{XY} = \rho = 0$.)

The correlation ratio, η, may be calculated from the data in the correlation table, Table 12.2, by the following operations:

$$\eta = \sqrt{\frac{\Sigma y_b^2}{\Sigma y_t^2}}$$

where Σy_b^2 is the between-categories (between-columns) sum-of-squares for the Y distribution, and Σy_t^2 is the total sum-of-squares for the Y distribution.

These sum-of-square terms, the Y distribution sum-of-squares, result from a sum-of-squares breakdown similar to that previously examined in analysis of variance. That is, the data can be conceived to be a two-variable table of classification with 9 rows and 15 columns, and one observation per rc cell. Thus the total sum-of-squares and the sum-of-squares for means (between-categories) are computed in the usual fashion.

196

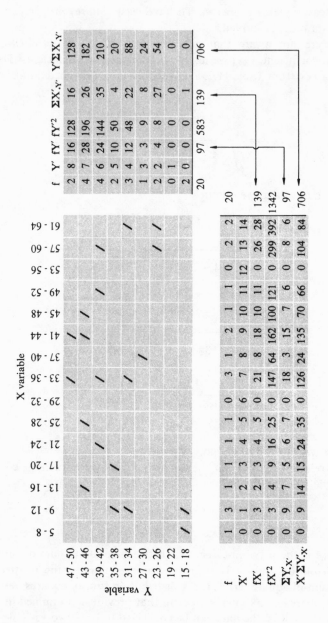

Right (upper) frequency table:

f	Y'	fY'	fY'²	ΣX'.ᵧ	Y'ΣX'.ᵧ
2	8	16	128	16	128
4	7	28	196	26	182
4	6	24	144	35	210
2	5	10	50	4	20
3	4	12	48	22	88
1	3	3	9	8	24
2	2	4	8	27	54
0	1	0	0	0	0
2	0	0	0	1	0
20		97	583	139	706

20   139   1342   97   706

X variable

Y variable

Bottom summary table (X columns 5-8 … 61-64):

	5-8	9-12	13-16	17-20	21-24	25-28	29-32	33-36	37-40	41-44	45-48	49-52	53-56	57-60	61-64
f	1	3	1	1	1	1	0	3	1	2	1	1	0	2	2
X'	0	1	2	3	4	5	6	7	8	9	10	11	12	13	14
fX'	0	3	2	3	4	5	0	21	8	18	10	11	0	26	28
fX'²	0	3	4	9	16	25	0	147	64	162	100	121	0	299	392
ΣY'.ₓ	0	9	7	5	6	7	0	18	3	15	7	6	0	8	6
X'ΣY'.ₓ	0	9	14	15	24	35	0	126	24	135	70	66	0	104	84

(Y variable rows: 47-50, 43-46, 39-42, 35-38, 31-34, 27-30, 23-26, 19-22, 15-18)

TABLE 12.2

The formulas below provide computational methods for the terms in the correlation table:

$$\Sigma y_b{}^2 = \Sigma \frac{(\Sigma Y')^2}{fX'} - \frac{\Sigma(\Sigma Y')^2}{n} = 550.5 - \frac{(97)^2}{20}$$

$$= 550.5 - \frac{9409}{20} = 550.5 - 470.45$$

$$= 80.05$$

$$\Sigma y_t{}^2 = \Sigma fY'^2 - \frac{(\Sigma fY')^2}{n} = 706 - 470.45$$

$$= 235.55$$

Column	fX	$\Sigma Y'_{.x}$	$(\Sigma Y')^2$	$(\Sigma Y')^2/fX$
0	1	0	0	0
1	3	9	81	27
2	1	7	49	49
3	1	5	25	25
4	1	6	36	36
5	1	7	49	49
6	0	0	0	0
7	3	18	324	108
8	1	3	9	9
9	2	15	225	112.5
10	1	7	49	49
11	1	6	36	36
12	0	0	0	0
13	2	8	64	32
14	2	6	36	18
				$\Sigma = 550.5$

TABLE 12.3

$$\eta = \sqrt{\frac{\Sigma y_b{}^2}{\Sigma y_t{}^2}}$$

Substituting, we obtain

$$\eta = \frac{80.05}{235.55}$$

$$= .34$$

Since η is not a product-moment correlation coefficient it cannot be interpreted directly. That is, the magnitude of the correlation ratio cannot be directly interpreted. However, if the correlation between the two variables is linear, η will be approximately equal to r_{XY} (η has no sign, i.e. it is always positive). If the data are curvilinearly related, η will be larger than r_{XY}. The discrepancy between these two indices is related to the extent of the departure from linearity of the relationship between the two distributions.

The significance of the obtained magnitude of η can be established, indirectly, with the F test. The total and the between- sum-of-squares have already been obtained. The within- sum-of-squares can also be obtained, as a residual. The data are summarized in a standard table:

$$\Sigma y_w{}^2 = \Sigma y_t{}^2 - \Sigma y_b{}^2$$
$$= 235.55 - 80.05$$
$$= 155.50$$

Summary table

Source	Sum-of-squares	df	Mean square
Between	80.05	14 (c − 1)	5.72
Within	155.50	6 (n − c)	25.92
Total	235.55	19 (n − 1)	

$$F = \frac{MS_b}{MS_w} = \frac{5.72}{25.92}$$
$$= .22$$

$P(F = .22)$, with 14df and 6df, is greater than .05. We conclude that the correlation ratio, η, is not significant.

The usefulness of an index such as η is that, unlike t and F, and like other correlation indices, it indicates type as well as degree of relationship.

The Tetrachoric Correlation Coefficient: r_t

The tetrachoric correlation coefficient, r_t, is relatively difficult to compute. For this reason, we will discuss a short estimate form of r_t. The tetrachoric correlation coefficient is applicable when two correlated variables, X and Y, are normally distributed and the regression between them is linear. When such data are categorized into two sets of dichotomies, r_t may be used to determine the relationship between the distributions.

This coefficient is similar to the biserial correlation coefficient, r_b, and like r_b and r_{XY}, r_t is a product-moment correlation coefficient. Thus, r_t can also be used to estimate r_{XY}. However, the use of r_t to estimate r_{XY} is desirable only when we are dealing with distributions consisting of large numbers of observations.

When both distributions are continuous and have been dichotomized, they can be summarized in a table:

Variable X

		Category A	Category B	
	Category a	a	b	a + b
Variable Y	*Category* b	c	d	c + d
		a + c	b + d	

If the two variables, X and Y, are continuous and normally distributed, and if the regression between the two is linear, then with the proper manipulation we should be able to specify what proportion of the observations fall into any sector of the correlation scattergram. If we reduce the scattergram to a diagram representing the proportion of cases in each quadrant, as in the table above, it should be relatively simple to specify what proportion of the X and Y variable observations falls into each quadrant.

It is important to note that, for distributions with relatively small N's, p should be approximately equal to q, where p represents the proportion of cases in the A or B category and q is equal to $1 - p$. When the departure approaches .85 and .15, for instance, the ratio becomes unstable. The approximation to r_t can be established by determining which ratio, $\frac{ad}{bc}$ or $\frac{bc}{ad}$, will provide a value greater than 1.00. That is, the larger product always goes into the numerator of the ratio. The quantitative value obtained from the ratio is, then, looked up in Table A.13, p. 280. This table gives approximate values of r_t for the obtained ratio.

For illustrative purposes, let us examine the following data. One hundred individuals were given two tests. Those who exceeded a certain criterion value on Test X obtained passing marks, while those who scored below that point failed. Similarly, those exceeding a criterion score on Test Y were classified as passing, while those who did not were classified as failing. The 100 individuals are classified in a table:

		Test X		
		passing	failing	
	passing	a 25	b 25	a + b 50
Test Y	failing	c 35	d 15	c + d 50
		a + c 60	b + d 40	100

$$\frac{ad}{bc} = \frac{(25)(15)}{(35)(25)} = \frac{375}{875} = .43$$

Since this ratio is smaller than 1.00,

$$\frac{bc}{ad} = \frac{(35)(25)}{(25)(15)} = \frac{875}{375} = 2.23$$

For the given table values, 2.33 falls between 2.29 and 2.34. Thus $r_t = .32$.

The ϕ Coefficient

When each variable is truly dichotomous, the relationship between the X and Y distribution can be determined with the ϕ coefficient (Greek phi).

The data are arranged in a table like the one on p. 199, that is, r_t. Then ϕ is defined as

$$\phi = \frac{(ab - bc)}{\sqrt{(a + b)(c + d)(a + c)(b + d)}}$$

To illustrate the computation of ϕ, let us examine the following data: 100 individuals were given two *situational* tests: passing or failing on the tests was determined in terms of whether a particular *appropriate* response was observed.

		Test X		
		passing	*failing*	
Test Y	*passing*	a 20	b 32	a + b 52
	failing	c 17	d 31	c + d 48
		a + c 37	b + d 63	100

$$\phi = \frac{(ad - bc)}{\sqrt{(a + b)(c + d)(a + c)(b + d)}}$$

$$= \frac{[(20)(31) - (32)(17)]}{\sqrt{(20 + 32)(17 + 31)(20 + 17)(32 + 31)}}$$

$$= \frac{(620 - 544)}{\sqrt{(52)(48)(37)(63)}}$$

$$= \frac{76}{\sqrt{5818176}} = \frac{76}{2412.09}$$

$$= .03$$

Rank-order Correlation Coefficients

Under certain circumstances, the X and Y pairs in correlated data are not observations falling on a continuous scale. That is, the observations may fall on a nominal scale. For instance, instead of giving the actual score obtained on a test, an individual may be assigned a rank based upon his performance in comparison with that of other persons who took that test. If, for instance, two tests, were administered to each of n individuals, their score may be expressed in ranks so that there would be one rank for Test X and another for Test Y. The problem in determining whether there is a relationship between the performance of the individuals on both tasks is to determine whether there is a relationship between the ranks. We must determine whether the ranks are consistent, or stable. The method appropriate for determining the stability of ranks is a test of independence of ranks.

Spearman's Rank-difference Correlation Coefficient: ρ_s

Spearman's rank-difference correlation coefficient, ρ_s, has been qualified with the subscript s (to indicate Spearman) so that it will not be confused with ρ, the product-moment correlation coefficient.

When observations are obtained in the form of ranks, it is appropriate to use ρ_s. It may also be used to estimate r_{XY} when the sample size is relatively small, e.g. n = 20 or less. That is, continuous observations may arbitrarily be converted to ranks and then the rank-difference correlation coefficient may be used.

The rank-difference correlation coefficient does not depend upon the validity of the parametric assumptions. Thus, when sample size is small—so small that it is extremely unlikely that the samples even approximate a normal distribution—ρ is an excellent estimate of r_{XY}.

The coefficient ρ_s may assume the same values as r_{XY}: its value may range from -1.00 (exactly opposite ranking) to 0.00, to $+1.00$ (perfect rank agreement). The significance of a particular value of ρ_s depends, in part, upon sample size (Table A.9).

Computing ρ_s

The bivariate distribution used to illustrate the computation of η will also be used to illustrate the computation of ρ_s. The observations in the respective variables are transformed to ranks by assigning the value 1 to the highest quantity, 2 to the next highest, and so on, finally assigning 20 to the lowest quantity (n = 20). The data consists of pairs of ranks based upon the original scores obtained by each of the persons who took the X Test and the Y Test. The distribution is given below.

It is not essential to indicate the sign of the difference $R_1 - R_2$, since that difference, D, is squared (D^2). To determine ρ_s, this computational formula may be used:

$$\rho_s = 1 - \frac{6\Sigma D^2}{n(n^2 - 1)}$$

where n = number of pairs.

Substituting our data, we obtain

$$1 - \frac{6(1138.50)}{20(400-1)} = 1 - \frac{6831}{7980}$$
$$= .14$$

Subject	Rank			Rank	$R_1 - R_2$	$(R_1 - R_2)^2$
	X	R_1	Y	R_2	D	D^2
1	5	20	16	20	0	0
2	9	19	33	15	4	16
3	19	15	38	11.5	3.5	12.25
4	13	16	43	5.5	10.5	110.25
5	27	13	45	3.5	9.5	90.25
6	21	14	41	7	7	49
7	35	9.5	40	8.5	1	1
8	12	17	36	13	4	16
9	35	9.5	47	1.5	8	64
10	41	8	43	5.5	2.5	6.25
11	50	5	40	8.5	−3.5	12.25
12	48	6	45	3.5	2.5	6.25
13	61	2	31	16	−14	196
14	33	11	34	14	−3	9
15	57	4	39	10	−6	36
16	58	3	26	17.5	−14.5	210.25
17	63	1	26	17.5	−16.5	272.25
18	28	12	38	11.5	0.5	.25
19	11	18	18	19	−1	1
20	43	7	47	1.5	5.5	30.25

$$\Sigma = 1138.50$$

The Significance of ρ_s

Since ρ_s is a product-moment correlation coefficient, it may usually be interpreted much as r_{XY} would be. It is probably as reliable as r_{XY}, even though its standard error, σ_{ρ_s}, is difficult to determine precisely.

When n is reasonably large, e.g. 25 or more, the standard error of ρ_s can be approximated:

$$\sigma_{\rho_s} = \frac{1}{\sqrt{n-1}}$$

When the sample size is small, a table may be used to determine the significance of ρ_s. Table A.9, p. 276, provides confidence limits for coefficients obtained from samples of varying size. Note that the table gives values for

r_{XY} and the population correlation coefficient, ρ. However, ρ_s may be sub-stituted.

When the sample size is greater than 10, t may be used. The formula for t is similar to the one previously encountered in connection with r_{XY}:

$$t = \frac{\rho_s}{\sqrt{1 - \rho_s{}^2}} \sqrt{n - 2}$$

The probability of the resultant t is obtained from table A.4, which gives the probability of t, with $n - 2$ df.

Since the sample size is 20, there are 18 degrees of freedom. Table A.8 indicates that a ρ value of .4438 is required to reject the null-hypothesis; that is, $\rho_s \neq 0$, at the 5 per cent level of significance. Thus we may conclude that our obtained ρ value of .09 is not significantly different from zero.

Kendall's Coefficient of Concordance: W

Frequently there are conditions that make it desirable to determine whether a particular trait or other behavior index is stable. For instance, there are observations of a qualitative nature that may describe a particular phenomenon: aesthetic judgments of artistic creations, a personality characteristic of an individual, the anxiety-arousing component of some stimuli, etc. The procedure that an experimental investigator might follow in order to establish the stability of such judgments is to ask several judges to assign rank-order values to a number of related categories. For instance, several paintings might be ranked with respect to a particular criterion, e.g. preference or balance. Or the concepts of words might be ranked with respect to the extent to which they are rated masculine or feminine (this was done in one particular study).

In order to determine whether such ranks actually follow a stable pattern, a comparison is made of the order in which the items are assigned rank value according to the ranking criterion. If the criterion is meaningful, and if the ranking procedure is stable, then most of the judges will assign approximately equal rank-order to any particular item.

One method of estimating the stability of ranking by more than two judges is *Kendall's coefficient of concordance*, W.

Table 12.4 illustrates the procedure. Four judges were required to rank 15 photographs of the faces of men with respect to the extent to which physiognomy reflects masculinity. The value 1 is assigned to the photograph held to represent the most masculinity, while the value 15 (the rank 15) is assigned to represent the least.

Ranks

Picture	judge 1	judge 2	judge 3	judge 4	Σ Ranks	$(\Sigma\Sigma R/n - \Sigma R)$ D	D^2
a	1	1	3	1	6	26	676
b	11	5	12	11	9	23	529
c	13	12	10	12	9	23	529
d	3	2	2	2	19	13	169
e	9	9	8	8	25	7	49
f	10	15	13	14	22	10	100
g	2	3	1	3	27	5	25
h	4	6	5	7	34	−2	4
i	7	4	4	4	37	−5	25
j	5	8	6	6	41	−9	81
k	12	10	9	10	47	−15	225
l	6	11	11	9	39	−7	49
m	8	7	7	5	55	−23	529
n	14	13	15	13	52	−20	400
o	15	14	14	15	58	−26	676

$\left(\Sigma\Sigma R/n = \dfrac{480}{15} = 32\right)$ $\Sigma\Sigma R = 480$ $\Sigma D^2 = 4066$

TABLE 12.4

The coefficient of concordance, W, is defined as

$$W = \frac{12\Sigma D^2}{(n_j)^2(n)(n^2 - 1)}$$

Substituting our data, we obtain

$$W = \frac{12(4066)}{(16)(15)(225 - 1)} = \frac{48792}{53760}$$

$$= .908, \quad \text{or} \quad .91$$

where n_j is the number of judges.

The coefficient of concordance, W, may range from zero to 1.00. Thus complete agreement by all judges would be represented by a coefficient of 1.00, while complete disagreement would result in a computed coefficient of 0.00.

To determine the significance of a particular value of W, Table A.11, p. 278, may be used. The column on the left of the table gives the number of judges, n_j. The row at the top gives the number of categories ranked, n. Thus, we would enter the last column (since 10 is the highest n given) at the second row ($n_j = 4$) at the .05 level. The table value is .48. We conclude that any obtained value of W that exceeds .48 is significantly different from zero at the .05 level of significance. Our empirically obtained W exceeded that table value.

Problems

A test was administered to a class of 100 students. The distribution of correct (X) and incorrect (Y) answers is given in terms of total test score, for one of the test items:

Test Score	X	Y
95 -	15	8
90 - 94	10	4
85 - 89	9	4
80 - 84	8	4
75 - 79	8	3
70 - 74	5	0
65 - 69	6	3
60 - 64	3	2
55 - 59	4	0
- 54	2	2

1. Compute r_b from the distributions above.
2. Employing the same data, compute r_{pb}.
3. Three judges ranked five finalists in a beauty contest. Determine the ρ_s coefficient between judge 1 and 3.

Finalist	Judges 1 2 3		
a	1	2	1
b	3	1	2
c	2	3	4
d	4	4	3
e	5	5	5

How consistent are these judges?

4. Employing the distributions in Table 11.1, compute the coefficient η.
5. Employing the judgment distributions in 3, above, compute W.

Answers

1. The first step in computing r_b is to determine the value of $\bar{X}_p$, $\bar{X}_T$, and S_X. These may be computed from a grouped frequency distribution method.

Test Score	X	Y	f_t*	X'	$f_t X'$	$f_t X'^2$	$f_X X'$
95 -	15	8	23	5	115	575	75
90 - 94	10	4	14	4	56	224	40
85 - 89	9	4	13	3	39	117	27
80 - 84	8	4	12	2	24	48	16
75 - 79	8	3	11	1	11	11	8
70 - 74	5	0	5	0	0	0	0
65 - 69	6	3	9	−1	−9	9	−6
60 - 64	3	2	5	−2	−10	20	−6
55 - 59	4	0	4	−3	−12	36	−12
- 54	2	2	4	−4	−16	64	−8
	$\Sigma = 70$	$\Sigma = 30$	$\Sigma = 100$		$\Sigma = 198$	$\Sigma = 1104$	$\Sigma = 134$

* f_t is the total frequency: $X + Y$.

$$\bar{X}_p = 72 + \frac{134}{70}(5)$$

$$= 72 + 1.91(5)$$

$$= 81.55$$

$$\bar{X}_T = 72 + \frac{198}{100}(5)$$

$$= 72 + 1.98(5)$$

$$= 81.90$$

$$S_X = \sqrt{\frac{1104 - \frac{(198)^2}{100}(5)^2}{100}}$$

$$= \sqrt{177.99}$$

$$= 13.34$$

and

$$p = \frac{70}{100}$$

$$= .70$$

Thus,

$$r_b = \frac{\bar{X}_p - \bar{X}_T}{S_X}\left(\frac{p}{Y}\right)$$

$$= \frac{81.55 - 81.90}{13.34}(2.01)$$

$$= -.05$$

2.

$$r_{pb} = \frac{\bar{X}_p - \bar{X}_T}{S_X}\left(\sqrt{\frac{p}{Y}}\right)$$

$$= \frac{81.55 - 81.90}{13.34}(1.42)$$

$$= -.01$$

3.

R_1	R_2	D	D^2	Finalist
1	1	0	0	a
3	2	1	1	b
2	4	2	4	c
4	3	1	1	d
5	5	0	0	e
			$\Sigma = 6$	

$$1 - \frac{6\Sigma D^2}{n(n^2 - 1)}$$

$$= 1 - \frac{6(36)}{5(24)}$$

$$= 1 - \frac{216}{120}$$

$$= -0.8$$

X variable

Y variable	26-27	28-29	30-31	32-33	34-35	36-37	38-39	40-41	42-43	44-45
44-45								/		
42-43										
40-41										
38-39							/			
36-37							/			
34-35										
32-33					/	/				
30-31										
28-29						//				
26-27	//		/							

	f	Y'	fY'	fY'²	ΣX'.Y'	Y'ΣX'.Y'
44-45	1	9	9	81	7	63
42-43	0	8	0	0	0	0
40-41	0	7	0	0	0	0
38-39	1	6	6	36	6	36
36-37	1	5	5	25	6	30
34-35	0	4	0	0	0	0
32-33	2	3	6	18	11	33
30-31	0	2	0	0	0	0
28-29	2	1	2	2	8	8
26-27	3	0	0	0	5	0
	10		28	162	43	170

f	0	2	0	1	2	1	3	1	0	0	10
X'	0	1	2	3	4	5	6	7	8	9	
fX'	0	2	0	3	8	5	18	7	0	0	43
fX'²	0	2	0	9	32	25	108	49	0	0	225
ΣY'.X'	0	0	0	0	2	3	14	9	0	0	28
X'ΣY'.X'	0	0	0	0	8	15	84	63	0	0	170

Column	fX	ΣY'.x	(ΣY')²	(ΣY')²/fX
0	0	0	0	0
1	2	0	0	0
2	0	0	0	0
3	1	0	0	0
4	2	2	4	2
5	1	3	9	9
6	3	14	196	65.3
7	1	9	81	81
8	0	0	0	0
9	0	0	0	0
	Σ = 10	Σ = 28	Σ = 290	Σ = 157.30

$$\Sigma y_b{}^2 = 157.30 - \frac{(28)^2}{10}$$
$$= 157.30 - 78.40$$
$$= 78.90$$

$$\Sigma y_t^2 = 170.00 - 78.40$$
$$= 91.60$$
$$\eta = \sqrt{\frac{\Sigma y_b^2}{\Sigma y_t^2}}$$
$$= \sqrt{\frac{78.90}{91.60}}$$
$$= \sqrt{.861}$$
$$= .93$$

5. Computation of W.

$$W = \frac{12\Sigma D^2}{(n_J)^2(n)(n^2 - 1)}$$

			Σ Ranks	D	D²
1	2	1	4	5	25
3	1	2	6	3	9
2	3	4	9	0	0
4	4	3	11	−2	4
5	5	5	15	−6	36
			ΣΣ = 45		Σ = 74

$$\frac{\Sigma\Sigma \text{ Ranks}}{n} = \frac{45}{5} = 9$$

$$W = \frac{12(74)}{(9)(5)(24)}$$

$$= \frac{888}{1080}$$

$$= .822$$

XIII

Non-parametric Statistics

In previous chapters we examined methods of analysis by which we compared two or more sample statistics obtained (presumably) from samples that statisfied the parametric assumptions. These statistical tests, of the difference between sample distributions and the inferences about the populations from which they were obtained, rely on *a priori* assumptions about the sampling methods.

There are useful distributions of observations whose form is *not* known, or whose form departs markedly from the normal distribution. Non-parametric tests do not rely on known proportions of the normal distribution for estimates of the probability of a given event E. Such tests are useful, therefore, when the parametric assumptions cannot be satisfied.

We will discuss some of the more commonly used non-parametric statistics in this chapter.

Chi square: χ^2

Let us suppose, hypothetically, that a particular population can be described by a variable consisting of k categories. Each of these categories represents a different kind of observation that can be made on the population. In addition, it is known that, in a given population, the frequency with which each category is represented is $F_1, F_2, F_3 \ldots, F_k$. Empirical observations of the population may result in the frequency distribution $f_1, f_2, f_3 \ldots, f_k$. It might be desirable to determine whether the empirically obtained frequency distribution is representative of the population.

If $F_1 = f_1$, $F_2 = f_2$, $F_3 = f_3, \ldots,$ and $F_k = f_k$, then the discrepancy between the empirically obtained distribution and the known population

211

distribution is zero, and the sample is an unbiased representation of the population.

However, empirical frequency distributions are seldom identical to population distribution. Thus the question is really how much difference between the empirical and population distribution we can tolerate while still holding to the assumption that the sample represents the population.

In order to test for the significance of the discrepancy between an empirical frequency distribution and a hypothetical frequency distribution (the hypothetical frequency distribution being the population distribution), the χ^2 test may be employed. Chi square is defined as follows:

$$\chi^2 = \cdot \Sigma \frac{(f - F)^2}{F}$$

where F is the population, or hypothetical frequency and f is the obtained, or empirical, frequency in each category of the variable. The resulting χ^2 has $k - 1$ df.

Table A.5(a), p. 268, provides α per cent limits of the sampling distribution of χ^2 for various df. This table may be employed to determine whether a given value of χ^2 exceeds what may be expected on the basis of chance α per cent of the time. The sampling distribution of χ^2 varies with df: it is skewed when df is relatively small, and symmetrical when df is relatively large (30 or more). But the sampling distribution of χ^2 is determined by $k - 1$ df and not by Σf, i.e. by categories, not by sample size. Consequently, it is suggested that none of the f's be less than 1 and not more than 20 per cent of the f's be less than 5.

Computing χ^2

The chi square test is applicable in those cases where the data consist of per cent, proportions, or distributions that can be readily converted to such values.

To illustrate the computation of χ^2, we shall refer to the hypothetical distribution of values that should be obtained when two dice are tossed. Figure 3.1, p. 42, provides the hypothetical distribution for N tosses of two dice. This distribution and an empirical distribution obtained by actually tossing two dice 36 times are shown in Table 13.1.

$$\chi^2 = \Sigma \frac{(f - F)^2}{F}$$

$$= 5.03$$

Category	Hypothetical proportion	Hypothetical frequency (F)	Observed frequency (f)	Observed proportion
2	.028	1	2	.056
3	.056	2	3	.083
4	.083	3	1	.028
5	.111	4	5	.139
6	.139	5	5	.139
7	.167	6	5	.139
8	.139	5	6	.167
9	.111	4	5	.139
10	.083	3	1	.028
11	.056	2	2	.056
12	.028	1	1	.028
	1.001*			1.002*

* Rounding off.

Distribution of hypothetical proportion and predicted frequency, and empirically observed proportion and frequency.

Category	F	f	$(f - F)^2$	$\dfrac{(f - F)^2}{F}$
2	1	2	1	1.00
3	2	3	1	.50
4	3	1	4	1.33
5	4	5	1	.25
6	5	5	0	0
7	6	5	1	.17
8	5	6	1	.20
9	4	5	1	.25
10	3	1	4	1.33
11	2	2	0	0
12	1	1	0	0

$$\Sigma = 5.03$$

TABLE 13.1

The probability that $X^2 = 5.03$, df $= 10$ (k $- 1$), on the basis of chance alone, is greater than 5 per cent. That is, the 95th percentile of the X^2 distribution for the given df has a limit equal to 18.31. In order to exceed the 95th percentile of the sampling distribution of X^2 for the given df, an obtained X^2 would have to have a value exceeding 18.31.

In the example above, the hypothetical frequency distribution was known. That is, the hypothetical distribution of values obtainable when two dice are thrown is described by the binomial distribution. There are instances when the hypothetical distribution is *not* known. In that case, the investigator may wish to stipulate what the hypothetical distribution *should* be like, according to some theory concerning the population from which the observations are obtained.

In order to illustrate how the hypothetical distribution may be generated. we will consider the following.

Your company, A, has introduced a new product, which, though considerably costlier than the products of three competitors, B, C, and D, is significantly better than all of them. After your product has been on the market for several months, you wish to evaluate sales in terms of the market as a whole to determine whether the product should be retained, its profit margin being smaller than you would like.

A survey of 1000 consumers has resulted in the following distribution:

Consumers purchasing brand:

A	B	C	D
350	225	225	200

What questions about this frequency distribution are appropriate? You might say, for instance, that if your product does not appear to be significantly better than that of your competitors it should sell in the same proportion as their products. That is, if there is nothing that distinguishes your product from any of the others, then all four should have the same proportion of total sales. This hypothesis might be tested this way:

	A	B	C	D
Observed frequency	350	225	225	200
Hypothetical frequency	250	250	250	250

$$\chi^2 = \Sigma \frac{(f - F)^2}{F}$$

$$= \frac{(350-250)^2}{250} + \frac{(225-250)^2}{250} + \frac{(225-250)^2}{250} + \frac{(200-250)^2}{250}$$

$$= 55.00$$

The probability of $\chi^2 = 55$, df $= 3$, is smaller than 5 per cent and you conclude that the proportion of actual sales is not equally distributed among the competitors.

On the other hand, you might decide that there is no point in keeping the product unless your firm is selling as much as all other competitors combined:

	A	B	C	D
Observed frequency	350	225 + 225 + 200		
Hypothetical frequency	500	500		

$$\chi^2 = (350 - 500)^2/500 + (650 - 500)^2/500$$

$$= 90.00$$

The probability of $\chi^2 = 90.00$, df $= 1$, is smaller than 5 per cent, and you conclude that your actual sales are not equal to half the total sales.

The chi square test is useful, then, when an empirical distribution is to be matched to a known hypothetical distribution, or when you must test several different hypotheses about the relationship between an empirical distribution and some properties of a population in which you are interested. In the chi square tests above categories of only one variable are used. Now let us consider chi square tests with more than one variable.

Testing Independence in a Two-way Classification Table

To illustrate a chi square test of *independence* of the variables in a two-way classification table, let us suppose that someone in your school undertook a student poll on attitudes toward self-government. A sample of 140 students was obtained. The sample consisted of 80 male and 60 female students. The sample gave three categories to chose from: "greater participation in making school policies," "present system is adequate," and "neutral or no opinion."

The two variables in the classification table, i.e. in the frequency distribution, are *attitude* and *subject sex*. That is, there may be differences in the frequency with which each category of opinion is represented, but it may be, also, that the differential frequency for males is different from that for females.

A chi square test of *independence* is a test of the contention that the distribution of one variable should be the same irrespective of the second variable. That is, if subject sex and attitude are *independent*, then the proportion of persons favoring greater participation in making school policies should be the same for male and female students.

The two-way classification table, or *contingency* table, is shown in Table 13.2.

	Male	Female	T
"Greater participation"	10	3	13
"Accept *status quo*"	65	50	115
"No opinion"	5	7	12
T	80	60	140

	Male	Female	T
"Greater participation"	10 (7.41)*	3 (5.59)*	13
"Accept *status quo*"	65(65.55)*	50(49.45)*	115
"No opinion"	5 (6.84)*	7 (5.16)*	12
T	80	60	140

* Is the hypothetical frequency, which is computed as follows:

$$\frac{80}{140}(13) = 7.41$$

$$\frac{80}{140}(115) = 65.55$$

$$\frac{80}{140}(12) = 6.84$$

$$\frac{60}{140}(13) = 5.59$$

$$\frac{60}{140}(115) = 49.45$$

$$\frac{60}{140}(12) = 5.16$$

TABLE 13.2

The hypothetical frequencies, and their computation, are shown also. Note that determination of the hypothetical frequency appropriate in each cell is based on the *margin totals:* for instance, if the variables are independent the same proportion of males and females should respond in the first, second, and third categories. There are 12 persons in the first category ("greater participation") and we should expect to find that $\frac{80}{160}$ (12) of them are males. Likewise, we should expect to find that $\frac{60}{160}$ (12) of them are females. The product of the hypothetical frequencies is equal to the margin total.

The chi square test of independence is computed as follows:

$$\chi^2 = \frac{(10.00 - 7.41)^2}{7.41} + \frac{(65.00 - 65.55)^2}{65.55} + \frac{(5.00 - 6.84)^2}{6.84}$$

$$+ \frac{(3.00 - 5.59)^2}{5.59} + \frac{(50.00 - 49.45)^2}{49.45} + \frac{(7.00 - 5.16)^2}{5.16}$$

$$= 3.28$$

The distribution of χ^2, in this case, has $(r - 1)(c - 1)$ df, where r and c stand for rows and columns, respectively. The probability of $\chi^2 = 3.28$, df $= 2$, is greater than 5 per cent. Thus we fail to reject the hypothesis that the variables are independent.

Correction for Continuity

When the classification table is so constituted that df $= 1$ for the calculated chi square, it is recommended that the value .5 be subtracted from the absolute value of each of the f $-$ F differences before squaring:

$$\chi^2 = \Sigma \frac{[(f - F) - .5]^2}{F}, \quad \text{when df} = 1.$$

This procedure assures a better approximation of the chi square distribution.

Testing Goodness of Fit

We have seen that proportions of the sampling distribution of χ^2 for the particular df can be used to test the significance of the difference between two frequency distributions. It is therefore possible to test the significance of category frequency discrepancies between any observed distribution and any hypothetical distribution whose form is known. The form of the hypothetical distribution must be known so that the frequency distribution for the population it represents can be reconstructed. In the illustrative case below, an empirically obtained distribution, i.e. the frequency distribution in Table 2.3, p. 19, is compared with the normal distribution, to see whether, assuming the distribution represents an unbiased sample, it could have come from such a distribution.

Thus the chi square test may be used to determine whether any obtained frequency distribution "fits" a known distribution model.

The distribution has a mean of 22.98, a standard deviation of 5.03 and an n of 100.

Interval	Interval upper limit X_u	x $(X_u - \bar{X})$	z x/S	Cum. prop. of Z	Cum. prop. of Z, (Σf)	Exp. freq. F	Obs. freq. f
11 - 12	12.5	−10.48	−2.08	.02	2	2	2
13 - 14	14.5	−8.48	−1.69	.04	4	2	4
15 - 16	16.5	−6.48	−1.29	.10	10	6	5
17 - 18	18.5	−4.48	−.89	.18	18	8	7
19 - 20	20.5	−2.48	−.49	.30	30	12	11
21 - 22	22.5	−.48	−.10	.46	46	16	16
23 - 24	24.5	+1.52	+.30	.62	62	16	19
25 - 26	26.5	+3.52	+.70	.76	76	14	13
27 - 28	28.5	+5.52	+1.10	.86	86	10	9
29 - 30	30.5	+7.52	+1.50	.93	93	7	6
31 - 32	32.5	+9.52	+1.89	.97	97	4	5
33 - 34	34.5	+11.52	+2.29	.99	99	2	2
35 - 36	36.5	+13.52	+2.69	1.00	100	1	1

TABLE 13.3

Let us examine Table 13.3 column by column.

First column: "Interval." The frequency distribution is *grouped* with an interval 2, i.e. i = 2.

Second column: "Interval upper limit (X_u)." This is self-explanatory. The upper limit of each interval appears in this column.

Third column: "x, i.e. $X_u - \bar{X}$." In this column, each of the upper limit values is subtracted from the mean of the distribution, 22.98, to obtain the deviation of the upper limit from the mean of the distribution.

Fourth column: "z, i.e. $\frac{x}{S}$." The upper limit of the intervals are transformed to z, standard scores, to obtain standard score equivalents by dividing them by S (5.03).

Fifth column: "Cumulative proportion of Z." In this column, the standard scores, z, are normalized, by determining proportions of the area of the normal distribution, proportions of Z, *below* successive z values in terms of their corresponding Z values. For instance, the first value at the top of the column, $z = -2.08$, is entered in the table for proportions of the normal distribution, Table A.1, p. 263, at $Z = -2.08$. To avoid interpolation, there being very little difference between $Z = -2.08$ and $Z = -2.10$, the Z values will, in this example, always be approximated to the nearest multiple of 10. The proportion of the area below $Z = -2.10$ is .0178, or .02, rounding off. This value is entered at the top of the column. Next, $z = -1.69$. The corresponding area below $Z = -1.70$ is .0446, or .04, and so on. (Notice, for example, that in the previous column the calculated value of z would ordinarily be about .99999. This value is converted to 1.00 for the sake of convenience, since we wish to include all observations ($\Sigma f = n$) in the distribution. You will recall that the normal distribution has an $N = \infty$.)

Sixth column: "Cumulative proportions of Z multiplied by n." Having converted the empirically obtained z distribution to proportions of the normal distribution, we now wish to determine what proportion of the hypothetical distribution would belong in each interval if the empirical distribution were, in fact, normally distributed. Thus, instead of assigning appropriate frequency to each interval category based on $N = \infty$, we assign frequencies according to empirical sample size, n.

This operation is accomplished by multiplying the sample size n by the cumulative proportion of Z. This procedure determines what proportion of the sample would belong in each category if the empirical distribution were, in fact, normal, with $\Sigma f = N = \infty$ rather than n = 100.

For instance, if .02, i.e. 2 per cent, of the sample belongs in the first interval category, and the sample size is 100, then there would be .02(100) observations at that point in the distribution. We find that in the next category there is a cumulative proportion of .04. Thus, .04(100) observations occur below the upper limit of that inverval category. In the last interval category, we find that the cumulative proportion of n is 1.0(100).

Seventh column: "Expected frequency." By successive subtraction, we find the frequency in each interval category. Beginning with the last category, i.e. 35 - 36, we find that 100 observations are distributed below the upper limit of this category. There are 99 observations distributed below the upper limit of the next interval category, i.e. 33 - 34. Thus, there must be 1 observation between the upper limits of these two categories, i.e. 35 - 36 and 33 - 34, respectively. We find, next, that 97 observations are distributed below the upper limit of the category 31 - 32. Since 99 observations were distributed below the category 33 - 34, there must be 2 observations between these two interval categories, i.e. 31 - 32 and 33 - 34, respectively.

Eighth column: "Observed frequency." In this column, the empirical frequency distribution is given for the respective interval categories.

The procedure so far can be summarized in this way: if the empirical distribution were in fact normally distributed, with mean, $\bar{X} = \mu$, and standard deviation, $S = \sigma$, then we could use the normal distribution, by converting the empirical distribution to Z values, to determine what proportion of the observations should fall into each interval category. This hypothetical distribution can then be compared to the actual, obtained distribution, using the chi square test to evaluate the probability that the resulting discrepancy between the two distributions is the result of chance factors one might expect from sampling variation.

The test procedure is given below.

Interval	Observ. freq. f	Expect. freq. F	Regrouped frequency f	F	$(f - F)$	$(f - F)^2$	$\dfrac{(f - F)^2}{F}$
11 - 12	2	2					
13 - 14	4	2					
15 - 16	5	6	11	10	+1	1	.10
17 - 18	7	8	7	8	−1	1	.12
19 - 20	11	12	11	12	+1	1	.08
21 - 22	16	16	16	16	0	0	0
23 - 24	19	16	19	16	+3	9	.56
25 - 26	13	14	13	14	−1	1	.07
27 - 28	9	10	9	10	−1	1	.10
29 - 30	6	7	6	7	+1	1	.14
31 - 32	5	4	8	7	+1	1	.14
33 - 34	2	2					
35 - 36	1	1					

$$\Sigma = 1.31$$

TABLE 13.4

The three upper and three lower category frequencies have been combined (regrouped) in order to avoid categories with n less than 5 in the observed frequency distribution. The three upper and three lower categories have been regrouped in the expected frequency distribution for the same reason.

The chi square is then relatively simple to compute by obtaining the sum of the squared difference between the observed and expected frequency, and dividing it by the expected frequency. This obtained value is 1.31.

The probability of $\chi^2 = 1.31$, df = 8, is greater than 5 per cent, and we conclude that the obtained distribution does not differ significantly from a normal distribution.

χ^2 Test in Two-by-two Tables

Let us suppose that the result of a survey consists of observations in two categories, in each of two variables. For instance, 20 students in your class were asked to answer *yes* or *no* to two questions: "Would you rate your instructor as *good* or *bad*?"; and, "Is the grade your instructor gave you, at the end of the semester, a fair estimate of your performance in the class?"

A test of independence in a two-by-two table can be simplified in this way (note that for purposes of computational simplicity, the contingency table is labeled with letters to indicate computation steps):

		Question I	
Question II		Good	Bad	Total
	Fair	A	B	A+B
	Unfair	C	D	C+D
	Total	A+C	B+D	A+B+C+D = n

Substituting our data,

		Question I	
Question II		Good	Bad	Total
	Fair	4	2	6
	Unfair	6	8	14
	Total	10	10	20

$$\chi^2 = \frac{\left(|AD - BC| - \frac{n}{2}\right)^2 (n)}{(A + B)(A + C)(B + D)(C + D)}$$

$$= \frac{\left(|(4)(8) - (2)(6)| - \frac{20}{2}\right)^2 (20)}{6 + 10 + 10 + 14}$$

$$= \frac{(|32 - 12| - 10)^2 (20)}{40}$$

$$= \frac{2000}{40}$$

$$= 50.0$$

This χ^2 is distributed with $(r - 1)(c - 1)$, or 1df. The probability that χ^2 = 50.0, df = 1, is smaller than 5 per cent, and we conclude that there is a relationship between the grade obtained and rating of the instructor.

Fisher Exact Probability Test

The Fisher exact probability test is useful when data can be classified in a two-by-two table and consist of discrete variable categories. To illustrate the use and computational procedures in this test, let us suppose that you are considering two groups of human subjects. Each of these groups has received a number of standard treatments. Now, the group is divided into subgroups: a control group receiving a placebo treatment (a placebo is an inert substance, often administered to a control group so that its members think they are receiving the same thing as the experimental group) and a group receiving an experimental treatment. Following the administration of the treatments, active and inert, you wish to determine whether any significant difference now exists between the two subgroups. The contingency table is outlined below.

| | | *Effectiveness of Treatment* | | |
		Yes	*No*	*Total*
Treatment	*Active*	A	B	A + B
	Inert	C	D	C + D
	Total	A + C	B + D	N

The probability of this distribution is given by P, where

$$P = \frac{(A + B)!\,(C + D)!\,(A + C)!\,(B + D)!}{N!\,A!\,B!\,C!\,D!}$$

Substituting our data, we obtain

| | | *Effectiveness of Treatment* | | |
		Yes	*No*	*Total*
Treatment	*Active*	3	2	5
	Inert	2	3	5
	Total	5	5	10

and

$$P = \frac{(5!)\,(5!)\,(5!)\,(5!)}{10!\,(3!)\,(2!)\,(2!)\,(3!)}$$

$$= \frac{(5\cdot4\cdot3\cdot2)^4}{10\cdot9\cdot8\cdot7\cdot6\cdot5\cdot4\cdot3\cdot2(3\cdot2)(2)(2)(3\cdot2)}$$

$$= \frac{207360000}{522117600}$$

$$= .397$$

As expressed in the contingency table, the probability of the relationship between frequency of success and failure of the treatment in the control group and the experimental group is approximately .4. Since such a frequency distribution, to be affected differentially by the treatment, would have to be highly improbable as a chance events, i.e. less than 5 per cent or 1 per cent, we fail, with $P = .4$, to reject a hypothesis that the treatment groups are independent.

The Sign Test

Frequently, when a distribution of pairs of observations is obtained, it is possible to use percentiles of the t distributions to determine whether the differences between the distributions are significant. We have examined the procedure for this kind of analysis in a previous section. It should be noted, however, that the use of the t statistic is appropriate only when the distributions were obtained from a normally distributed population.

Occasionally, pairs of observations, even sets of pairs of observations, are obtained under widely varying conditions. The significance of the differences between the pairs obtained can be determined by the use of the sign test.

Let us suppose that n sets of pairs have been obtained. These pairs represent observations of an X variable and a Y variable. The differences between these pairs of observations would be

$$(X_1 - Y_1),\ (X_2 - Y_2),\ (X_3 - Y_3)\ldots,\ (X_n - Y_n).$$

If the differences between these pairs of observations were assigned $+$ when X is greater than Y, and $-$ when X is smaller than Y, then, on the basis of chance, approximately half the differences should be $+$, and the other half $-$.

The sign test is based on the hypothesis that there are an equal number of positive and negative differences. To illustrate computational procedures, let us examine the distribution of pairs of observations below.

X	Y	Difference
15	20	—
30	28	+
12	12	0
11	10	+
14	12	+
16	15	+
11	13	—
16	12	+
14	15	—
17	18	—

n = 10

In this table there are ten pairs of observations. Five pairs result in a positive difference, four pairs result in a negative difference, and one pair has a difference of 0 (neither positive nor negative). When the difference between a pair of observations is zero, that pair is dropped from the distribution. Thus, nine pairs remain: five positive and four negative differences.

The term s_i is used to indicate the number of times the less frequent sign is observed in the distribution. In our example, $s_i = 4$. When p = q = .5, it is possible to determine the probability of any number of signs, + or —, for a given number of pairs by use of the binomial probability distribution. But the simpler method is to note the critical value of s_i for a given sample size, n, in Table A.14.

In the example above, $s_i = 4$, n = 9. The critical value in the table is 1, for α = 5 per cent. An obtained value of s_i less than or equal to the critical value is significant at the given α level. The obtained value, 4, results in a failure to reject the hypothesis that the distribution of + and — signs differs significantly from equality.

The Mann-Whitney U Test

The Mann-Whitney U test is a simple test of group differences, based on relatively few assumptions. It is similar, in some ways, to the sign test.

The assumptions underlying its use are that the two samples compared are independently obtained samples, and that these samples come from populations whose variable is continuously distributed.

The computational procedure is simple and rapid. Observations in the two distributions, X and Y, are combined and arranged in one distribution, with values in ascending order of magnitude. It is not necessary for the samples to have the same size. The combined ranked X and Y distribution maintains the identity of the observations regarding the distribution from which each came. This procedure is illustrated in Table 13.5.

X	Y	X and Y in ascending order of magnitude	
35	13	35	X
20	27	35	X
29	14	33	Y
10	19	32	Y
22	32	30	X
21	15	29	Y
11	33	28	Y
35	28	27	Y
12	26	26	Y
16	30	22	X
		21	X
		20	X
		19	Y
		16	X
		15	Y
		14	Y
		13	Y
		12	X
		11	X
		10	X

TABLE 13.5

The test statistic U is the sum of the number of observations in one distribution preceding observations in the other distribution. In Table 13.6, U is determined for Y observations preceding X observations.

X and Y in ascending order of magnitude	Number of Y's preceding each X
35 X	10. . . . .
35 X	10. . . . .
33 Y	Y Y
32 Y	Y Y
30 Y	Y Y
29 X	7. . . . .
28 Y	Y Y Y
27 Y	Y Y Y
26 Y	Y Y Y
22 X	4. . . . .
21 X	4. . . . . .
20 X	4. . . . .
19 Y	Y Y Y Y Y Y
16 X	3. . .
15 Y	Y Y Y Y Y Y Y
14 Y	Y Y Y Y Y Y Y
13 Y	Y Y Y Y Y Y Y
12 X	0
11 X	0
10 X	0

$$T = 42$$
$$U = 0 + 0 + 0 + 3 + 4 + 4 + 4 + 7 + 10 + 10$$
$$= 42$$

TABLE 13.6

One cannot determine, usually, whether it is best to begin with Y values preceding X values, or X values preceding Y values. The simplest procedure is to obtain the smallest value for U. This can be done by performing the procedure for X preceding Y and that for Y preceding X, as shown in Table 13.7. The simplest way to determine the proper value of U is to determine the alternate value of U, U', where

$$U' = (n_X)(n_Y) - U$$

and substitute the obtained values,

$$U' = 3 + 3 + 3 + 4 + 7 + 7 + 7 + 8 + 8 + 8$$
$$= 58$$

The 5 per cent values of U are given in Table A.12.

X and Y in ascending order of magnitude		Y's preceding X's	X's preceding Y's
35	X	10	
35	X	10	
33	Y		8
32	Y		8
30	Y		8
29	X	7	
28	Y		7
27	Y		7
26	Y		7
22	X	4	
21	X	4	
20	X	4	
19	Y		4
16	X	3	
15	Y		3
14	Y		3
13	Y		3
12	X		
11	X		
10	X		
		T = 42	T = 58

TABLE 13.7

Problems

1. In a hypothetical study of the effects of two different doses of a drug on the performance of a psychomotor tracking task, the following was found:

	dose 1	dose 2
No change	83	52
Impaired	180	150

Is there a significant difference between the two dose groups?

2. A recent (hypothetical) poll showed the following results relating income to education:

	below 5000	5001 to 7500	7501 to 10,000	above 10,001
College	5	60	125	8
High school	20	45	45	10
Grade school	65	75	20	2

Is there a significant difference between the categories in these two variables?

3. The following 15 pairs of observations is an observed distribution. Test the significance of the differences between the pairs in the distribution, employing the sign test (use the XY columns on p. 187).

Answers

1. Testing the significance of the difference between dose groups:

	dose 1		dose 2			
No change	83	A	52	B	A + B	135
Impaired	180	C	150	D	C + D	330
	263	A + C	202	B + D		n = 465

$$\chi^2 = \frac{n\left(|AD - BC| - \dfrac{n}{2}\right)^2}{(A + B)(C + D)(A + C)(B + D)}$$

$$= \frac{465\left(|83(150) - 52(180)| - \dfrac{465}{2}\right)^2}{(135)(330)(263)(202)}$$

$$= \frac{3796867406.25}{2366763300.00}$$

$$= 1.60$$

$P(\chi^2 = 1.60)$, df $= 1$, is greater than 5 per cent.

2. Determining the relationship between salary and income categories:

	below 5000	5001 to 7500	7501 to 10,000	above 10,001	T
College	37.03*	74.25*	78.41*	7.92*	
	5.00	60.00	125.00	8.00	198
High School	22.44*	45.00*	47.52*	4.80*	
	20.00	45.00	45.00	10.00	120
Grade school	30.29*	61.12*	64.15*	6.48*	
	65.00	75.00	20.00	2.00	162
T	90.00	180.00	190.00	20.00	

(* Represents hypothetical frequencies.)

$$\chi^2 = \frac{(5.00 - 37.03)^2}{37.03} + \frac{(20.00 - 22.44)^2}{22.44} + \frac{(65.00 - 30.29)^2}{30.29} \cdots$$

$$+ \frac{(2.00 - 6.48)^2}{6.48}$$

$$= 27.71 + 0.27 + 39.77 \ldots + 3.10$$

$$= 140.56$$

$P(\chi^2 = 140.56)$, df $= 6$, is less than 5 per cent.

3. Testing the significance of the differences between pairs in the given distribution:

X	Y	X − Y
47	35	+
37	35	+
34	31	+
29	29	0
27	26	+
25	24	+
25	22	+
23	22	+
23	21	+
22	20	+
21	19	+
21	16	+
18	16	+
17	16	+
13	15	−
11	14	−
8	11	−
7	10	−
6	9	−
1	1	0

$S_i = 5 \qquad n = 18$

$P(S_i = 5)$, $n = 18$ is greater than 10 per cent.

XIV

Tests for Trends

In previous chapters we have been concerned with tests of the significance of differences between the sample statistics of k populations. The test statistics were used to evaluate the magnitude of the differences in terms of the known proportions of the appropriate sampling distributions.

Generally, the differences between two or more sample statistics (reflecting the difference between two or more populations) are not considered to indicate differences in *trends* unless the different treatments constitute an *ordered variable*. The independent variable in a study is ordered when the different treatment categories are different quantities in one dimension. For instance, if the independent variable in a study consists of three different methods of instructing a given grade-school level of arithmetic, the dependent variable consists of test scores of three classes to which one method was applied. The different class test performance means, $\bar{X}$, do not constitute the basis for a trend. Different methods of instruction do not constitute an ordered variable. A scale of different methods of instruction is nominal only.

Suppose, however, that three groups of animal subjects are given a different quantity of food reinforcement for completing a maze-running task. Let us say that group I (the control group) receives the quantity Q, group II receives the quantity $Q + \Delta Q$, and group III receives the quantity $Q + 2\Delta Q$. This type of independent variable is ordered. The basis for tests of trends is the assumption that differences in the independent variable should result in *proportional* differences in the dependent variable.

Thus, a trend is understood to be a pattern reflecting the relationship between ordered independent and dependent variables. The category-scale relationship of the independent variable is ordinarily assumed on an *a priori* basis, and the scale of the obtained category statistics—i.e. the relationship

230

between category means—is tested to determine whether a trend can be inferred. The analysis of variance provides a useful method for establishing the presence of a trend when the assumption stated above fits the relationship between categories of the treatment groups.

One-variable Analysis of Variance: Test for Presence of a Trend

In an analysis of variance design that tests the significance of the difference between estimates of k populations, proportional relationships may be observed between treatment quantity and resulting category mean differences. The observed proportional differences may be the basis for tests to determine whether (1) treatment effectiveness is statistically significant (presence of a trend), and (2) category changes are indeed proportional to treatment quantities (the trend is linear).

It is also possible to test the parallelism of sets of trends, but here we will be concerned only with the first statement: there is a trend. To illustrate computational procedures, let us assume that the distribution in Table 14.1(a) consists of five groups of five observations each, on variable I, whose five

Variable

A	B	C	D	E
47	39	28	23	19
44	35	35	30	14
32	18	17	16	13
19	14	13	12	12
66	40	29	25	11

TABLE 14.1(a)

	A	A²	B	B²	C	C²	D	D²	E	E²
	47	2209	39	1521	28	784	23	529	19	361
	44	1936	35	1225	35	1225	30	900	14	196
	32	1024	18	324	17	289	16	256	13	169
	19	361	14	196	13	169	12	144	12	144
	66	4356	40	1600	29	841	25	625	11	121
T =	208	9886	146	4866	122	3308	106	2454	69	991
X̄ =	41.6		29.2		24.4		21.2		13.8	

$$\Sigma x_b^2 = 19092.20 - 16978.08 = 2114.12$$
$$\Sigma x_t^2 = 21605.00 - 16978.08 = 4626.92$$
$$\Sigma x_w^2 = 4626.92 - 2114.12 = 2512.80$$

TABLE 14.1(b)

Summary table

Source	Sum-of-squares	df	Mean square
Between	2114.12	4	528.53
Within	2512.80	20	125.64
Total	4626.92	24	

TABLE 14.1(b) (Continued)

treatment categories A, B, C, D, and E are given proportional quantities of a particular treatment.

In Figure 14.1, the mean of each category is plotted with regard to its appropriate treatment quantity. A line is drawn to connect the plotted points. This line represents the trend.

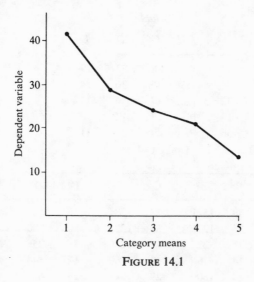

FIGURE 14.1

The computations necessary for a test of the presence of a trend are given in Table 14.1(b). You may wish to verify these computations.

The hypothesis that treatments are not differentially effective, i.e. that there is no trend, is tested by

$$F = \frac{MS_b}{MS_w}$$

Substituting the data in the summary table,

$$F = \frac{528.53}{125.64}$$

$$= 4.21$$

The probability that $F = 4.21$, $df_1 = 4$, $df_2 = 20$, is less than 5 per cent and we reject the hypothesis that the trend is not significant.

Comparison of Trends in a Two-variable Analysis of Variance

Under certain conditions, it is desirable to determine whether the trend indicated by the distribution of variable I category means is the same as the trend of each of several variable II treatment classifications. In a two-way classification table, it is possible that the variable I treatments are ordered, but it is not necessary that the variable II treatments be ordered also in order to apply trend analysis.

For instance, it might be desirable to determine whether five quantitatively different treatments result in similar trends when applied to three different types of subjects. An investigator might wish to determine whether the effect of different quantities of a drug agent on subjects performing a lever-pressing task selectively affect the response rate of two distinct genetic varieties of subjects. It may turn out that different quantities of the treatment variable do not affect all types of subjects in the same way.

To illustrate the procedure in the comparison of trends, the data in Table 14.2, plotted in Figure 14.2, will be employed. Note that category a of the variable II classification consists of the same data employed in the previous example, and category b in the variable II classification is an additional set of observations.

		Variable I				
		A	*B*	*C*	*D*	*E*
		47	39	28	23	19
		44	35	35	30	14
	a	32	18	17	16	13
		19	14	13	12	12
		66	40	29	25	11
Variable II						
		45	47	45	44	45
		42	35	39	40	42
	b	32	28	28	27	28
		44	38	42	41	40
		37	38	40	28	29

TABLE 14.2(a)

	A	A²	B	B²	C	C²	D	D²	E	E²
	47	2209	39	1521	28	784	23	529	19	361
	44	1936	35	1225	35	1225	30	900	14	196
a	32	1024	18	324	17	289	16	256	13	169
	19	361	14	196	13	169	12	144	12	144
	66	4356	40	1600	29	841	25	625	11	121

T =	208	9886	146	4866	122	3308	106	2454	69	991
$\bar{X}$ =	41.6		29.2		24.4		21.2		13.8	

	A	A²	B	B²	C	C²	D	D²	E	E²
	45	2025	47	2209	45	2025	44	1936	45	2025
	42	1764	35	1225	39	1521	40	1600	42	1764
b	32	1024	28	784	28	784	27	729	28	784
	44	1936	38	1444	42	1764	41	1681	40	1600
	37	1369	38	1444	40	1600	28	784	29	841

T =	200	8181	186	7106	194	7694	180	6730	184	7014
$\bar{X}$ =	40.0		37.2		38.8		36.0		36.8	

TT = 408   332   316   286   253

$$T_{++} = 1595$$

$\Sigma x_t^2 = 58230.00 - 50880.50 = 7349.50$

$\Sigma x_{\text{sub total}}^2 = 54789.80 - 50880.50 = 3909.30; \ df = rc - 1 = 9$

$\Sigma x_r^2 = 52597.75 - 50880.50 = 1717.25; \ df = r - 1 = 1$

$\Sigma x_c^2 = 52234.90 - 50880.50 = 1354.40; \ df = c - 1 = 4$

$\Sigma x_{\text{int}}^2 = 3909.30 - 1717.25 - 1354.40 = 837.65; \ df = (c - 1)(r - 1) = 4$

$\Sigma x_w^2 = 7349.50 - 3909.30 = 3440.20; \ df = nrc - rc = 40$

TABLE 14.2(b)

Summary table

Source	Sum-of-squares	df	Mean square
Column means	1354.40	4	338.60
Row means	1717.25	1	1717.25
Interaction (rows × columns)	837.65	4	209.41
Subtotal (cells)	3909.30	9	434.37
Within	3440.20	40	86.00
Total	7349.50		

TABLE 14.2(c)

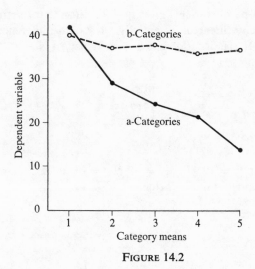

FIGURE 14.2

This classification table is similar to the one previously encountered in connection with two-way analysis of variance. A test for trend similarity—or trend difference, depending upon the predilection of the investigator—is essentially a test to determine whether the a category means trend is parallel to the b category means trend. In other words, irrespective of the linearity of the trends, are the two trends similar or different?

The test of parallel trends is

$$F = \frac{MS_{int}}{MS_w}$$

$$= \frac{214.16}{86.00}$$

$$= 2.49$$

The probability of $F = 2.49$, $df_1 = 4$, $df_2 = 40$, is less than 5 per cent, and we conclude that the trends of the a and b category means differ sufficiently to warrant the rejection of the hypothesis that they are parallel.

It is important to note that the application of analysis of variance to trend analysis depends upon the same assumptions encountered previously in connection with the analysis of variance. These assumptions specify population distribution normality, homogeneity of variances, independence of population means and variances, and random selection of category observations.

Finally, the test of parallel trends gives no indication that the trends coincide. A test of trend coincidence is, essentially, a test of the differences between the various variable II classification categories. This test is

$$F = \frac{MS_r}{MS_w}$$

$$= \frac{1717.25}{86.00}$$

$$= 19.97$$

The probability that $F = 19.97$, $df_1 = 4$, $df_2 = 40$, is less than .05. We conclude that the trends do not coincide.

XV

Trends in Time-domain Data Distributions

Time-domain data distributions consist of observations of an amplitude or a frequency function distributed in time. There are a number of populations of such observations that are concerned with one or both of these functions. Among these are blood pressure, electrical and tonus condition of musculature (EMG), cortical activity (EEG), precontractual-electrical cardiac activity (EKG), body fluid pressure and chemical composition changes, and changes in resistance of skin tissue to an electric current (GSR).

It is beyond the scope of the present text to examine any but the most elementary methods for studying such distributions and, in fact, it is not current practice to introduce these methods at the undergraduate college level. However, there has been, in recent years, an increase in the requirements of technical and academic skills of students in the undergraduate laboratory. This increase reflects advances in laboratory technology, in turn demanding of the student that he make observations on populations whose characteristics are not always clear: populations of analog measures, sequentially dependent, with unknown distribution and parametric measures.

In general, there are a few basic guidelines that it is well to bear in mind. Given an amplitude function distributed in time, a theorem in information theory stipulates that when sampling *every so often*, complete information may be obtained about the time function. The lowest value of *every so often* can be determined by considering the following: if a function contains no frequencies greater than C cycles per second, that function may be completely determined by sampling ordinates (in this case amplitude) at points $\dfrac{1}{2C}$ seconds apart through the complete time span of the function.

It has been pointed out that certain types of populations consist of events

237

dependent upon those just preceding them. This is at least one characteristic of most time-domain distributions: certain distributions consists of sequentially dependent observations—although this is not invariably so. Both the form of the distribution and the degree of dependence of the observations must be considered before parametric tests may be applied for comparison or inference purposes. The scope of this chapter precludes a definitive exposition of the various factors involved in the application of tests of significance to time-domain data distributions. A more detailed discussion can be obtained in the articles by Bradley, Haggard, Lacey, and Lathrop, cited in the Bibliography.

Analog Measures

In previous chapters we examined distributions of observations also described by continuous scales. However, the events were recorded as discrete digits. That is, the quantity of the scale category corresponding to the magnitude of the event, at a given point in time, was the observation. There are certain events whose magnitude is continuously in flux. These events, for the sake of convenience, are often recorded or observed as analogs of the event proper. In the life and behavior sciences, the analog of the event observed is often an electric current whose properties are measurable. This is not the only kind of analog, but it is the most frequently used. Two such measures, the galvanic skin response (GSR) and cardiac response (EKG) will be examined below to illustrate some of the more common transformations.

To observe the GSR, the investigator passes a small current between two points on the surface of the skin of the subject. A standard *resistance bridge* is used to measure changes in the resistance of the skin tissue between the contact, or electrode, surfaces. The proportional relationship between resistance and voltage, expressed by Ohm's Law $\left(E = \dfrac{I}{R} \right)$, indicates changes in voltage in terms of the changes in tissue resistance. The change in voltage or in current may be used to calibrate a scale of resistance values on a continuously moving paper chart on which change is recorded as an ink trace. The EKG is based on the amplification of electric currents generated by heart tissue, recorded on a strip chart recorder.

Irrespective of scale categories, populations of successive values of such continuous analog measures are not independent measures. The analog measure at any point in time is, by necessity, directly dependent on the one that preceded it. For any point on the analog trace, values are not free to vary randomly, or to assume any value at random.

The measures of EKG and GSR are often employed in the life and behavior sciences as indices of changes in the *arousal* states of the subject under investigation. The meaning and significance of arousal are not relevant here. But since these measures are frequently observed, we will examine methods for their analysis.

The analysis of events in the time domain, usually analog events, is often a trend analysis based on parametric methods employing percentiles of the F distributions. It should be pointed out that the transformation of analog measures to digital quantities by time sampling every t seconds is an attempt to transform the distributions so that standard statistical methods may be employed in data reduction and population comparisons.

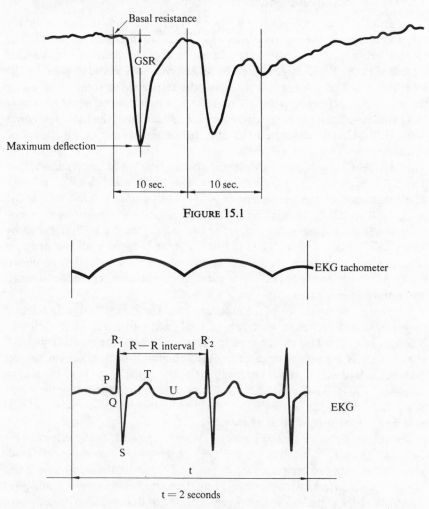

FIGURE 15.1

FIGURE 15.2

Transformations of GSR and EKG are discussed below. The methods suggested are those employed by investigators in the field, though there is some conjecture as to the most appropriate methods to employ in each case, and the issue is far from settled.

Characteristic forms of the GSR and EKG traces are shown in Figure 15.1 and Figure 15.2, respectively.

The Galvanic Skin Response

The galvanic skin response (GSR) is observed as a sudden and relatively sharp decrease in the resistance of skin tissue to the flow of an electric current between two locations on the surface. (This current, unlike that measured in the EKG, is generated by an external source and applied by the investigator.) The investigator measures the resistance at some point along the trace: the reference point. This reference point is the *basal resistance* (BR) and the magnitude of the decrease in resistance from the basal resistance is the GSR. Certain characteristics of distributions of such events should be considered.

In different subjects, as in different species, the basal resistance varies. The BR is different in organisms at rest and under conditions of arousal. The magnitude of the maximum resistance decrease is related to the BR level, GSR = f(BR), and is complicated by additional *somatic homeostatic mechanisms*, which have been described in some detail by Wilder and by Lacey (see Bibliography). Thus the investigator is faced with the problem of fitting to a parametric model a distribution of events for which no norms exist and which are confused by sequential dependency, correlated trends, and non-randomness.

Various transformations of the measures have been employed in an attempt to generate a distribution that can be analyzed with parametric methods. The data in Figure 15.3 can be used to illustrate some of the transformations. Figure 15.3 is an actual trace of a GSR obtained from a human subject under standard conditions. The ordinate scale in the trace is calibrated in K ohms (1000 ohms) resistance, and the observations are initially transformed by dividing by K, i.e. 1000. This initial transformation makes calibration and subsequent summaries easier.

A minimum change of 100 ohms (0.1K) is acceptable as the criterion for determining that a *signal* has actually been observed in a *recording*. (Recordings of this kind are characterized by *noise*, i.e. random small fluctuations. Noise often interferes with the actual signal and makes observation difficult.) In addition, the signal must have been observed within a finite time after the BR has been noted: the criterion for observing a GSR is that deflection must

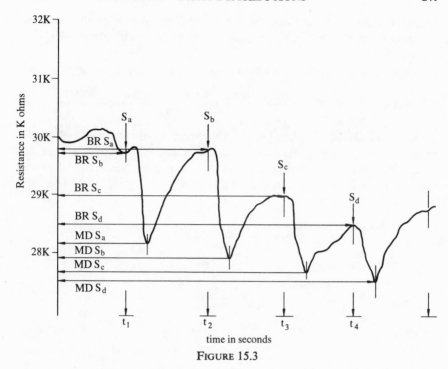

FIGURE 15.3

reach criterion magnitude between 0.5 seconds and 8.0 seconds from the BR reference point.

Thus the initial transformation consists of noting the criterion GSR deflection following the basal resistance reference point within the specified time span.

Change in Magnitude and Per Cent Change

To determine mean group trends, it would be of no value simply to super-impose the traces for the members of the groups. But individual and group trends can be shown by employing the magnitude of the GSR, i.e. the change in resistance, or by converting this change to a per cent change value.

In the transformations below we use the data in the trace shown in Figure 15.3. Group trends are simply the average of the transformed values.

The points S_a, S_b, S_c, and S_d represent environmental events (the independent variable) that, it is postulated, will result in the observation of the GSR (the dependent variable).

Figure 15.4 shows the decreasing trend in the BR pattern on successive occasions. However, since we are primarily interested in the pattern of successive differences between the BR and the magnitude of the resistance decrease from the BR, we can plot this change magnitude (GSR) as in Figure 15.5.

The magnitude of the decrease can also be converted to per cent change:

$$\text{Per cent change} = \frac{\text{BR} - \text{Resistance @ Maximum deflection}}{\text{BR}}$$

The per cent GSR is simpler to deal with when the ordinate scale units have a relatively large range. In the case of per cent change there can be at most 100 categories, since the decrease in resistance cannot be greater in magnitude than the basal level—i.e. there can be no negative resistance level. The ohms

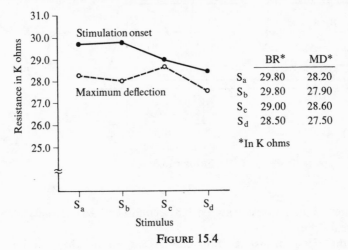

	BR*	MD*
S_a	29.80	28.20
S_b	29.80	27.90
S_c	29.00	28.60
S_d	28.50	27.50

*In K ohms

FIGURE 15.4

	BR	MD	BR-MD	% Change $\left(\dfrac{\text{BR-MD}}{\text{BR}}\right)$
S_a	29.80	28.20	1.60	5.4
S_b	29.80	27.90	1.90	6.5
S_c	29.00	28.60	.40	1.4
S_d	28.50	27.50	1.00	3.5

[All values are in K ohms]

FIGURE 15.5

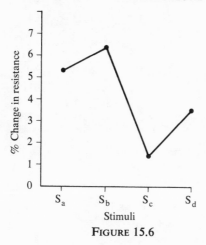

Stimuli

FIGURE 15.6

change to per cent ohms change transformation cannot be analyzed appropriately by parametric methods.

Conductance Magnitude and Per Cent Conductance Change

The GSR is defined above in terms of resistance change. It can also be defined as *conductance* change. The conductance is the reciprocal of the resistance.

$$\text{Conductance} = \frac{1}{\text{Resistance}}$$

The unit of conductance is called the *mho* (ohm spelled backward). GSR observations are usually expressed in millimhos. The distribution of events in Figure 15.3 can be converted to conductance units and the change and per cent change can be plotted as in Figures 15.7, 15.8, and 15.9.

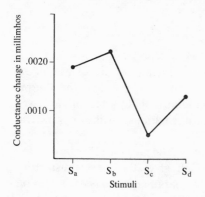

Stimuli

	BR	$\frac{1}{\text{BR}}$	MD	$\frac{1}{\text{MD}}$	$\left(\frac{1}{\text{BR}}\right) - \left(\frac{1}{\text{MD}}\right)^*$
S_a	29.80	.0336	28.20	.0355	—.0019
S_b	29.80	.0336	27.90	.0358	—.0022
S_c	29.00	.0345	28.60	.0350	—.0005
S_d	28.50	.0351	27.50	.0364	—.0013

[Since resistance is in K ohms, conductance is in millimhos.]

*Only absolute value is plotted.

FIGURE 15.7

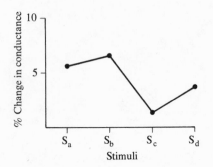

	$\dfrac{1}{BR}$	$\dfrac{1}{MD}$	$\dfrac{\left(\dfrac{1}{BR}\right)-\left(\dfrac{1}{MD}\right)}{1/BR}$
S_a	.0336	.0355	.057
S_b	.0336	.0358	.065
S_c	.0345	.0350	.014
S_d	.0351	.0364	.037

FIGURE 15.8

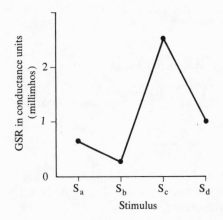

	BR	MD	$\dfrac{1}{BR-MD}$
S_a	29.80	28.20	.625
S_b	29.80	27.90	.154
S_c	29.00	28.60	2.500*
S_d	28.50	27.50	1.000

[These values are in K ohms and the decimal would move three places in actual ohms value. The transformation X/1000 is useful in that it reduces the size of the values but does not alter the relationship between them.]

FIGURE 15.9

The transformations presented so far have varying degrees of usefulness for summarizing the relationship between the observed baseline and the response consisting of a deflection from the baseline (BR) to a maximum lower value (MD). However, none of these transformations justifies parametric analysis of groups of such observations.

It has been suggested by some investigators in the area (see Haggard in the Bibliography) that one transformation which gives the most satisfactory distribution with respect to requirements for parametric analyses is the log transformation of the resistance change: log BR − MD.

Logarithmic Transformation

A logarithm is the exponent to which a value is raised in reference to a particular base quantity. Common logarithms have a base of 10. For the student who needs to refresh his memory, the table below, with the arbitrarily chosen value 2.58, will provide a rapid review. A table of logs for integers from 1 to 999 is given in the Appendix (Table A.17).

Antilogarithm	*Characteristic*	+	*Mantissa = Log*
$25,800.00 = 2.58 \ (10^4)$	4	+	$0.4116 = 4.4116$
$2,580.00 = 2.58 \ (10^3)$	3	+	$0.4116 = 3.4116$
$258.00 = 2.58 \ (10^2)$	2	+	$0.4116 = 2.4116$
$25.80 = 2.58 \ (10^1)$	1	+	$0.4116 = 1.4116$
$2.58 = 2.58 \ (10^0)$	0	+	$0.4116 = 0.4116$
$.258 = 2.58 \ (10^{-1})$	-1	+	$0.4116 = -1.4116$
$.0258 = 2.58 \ (10^{-2})$	-2	+	$0.4116 = -2.4116$
$.00258 = 2.58 \ (10^{-3})$	-3	+	$0.4116 = -3.4116$
$.000258 = 2.58 \ (10^{-4})$	-4	+	$0.4116 = -4.4116$

The distribution of log resistance change is shown below.

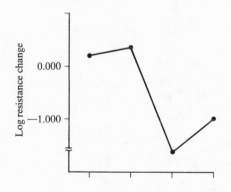

	BR	MD	BR-MD	Log BR-MD
S_a	29.80	28.20	1.60	0.204
S_b	29.80	27.90	1.90	0.279
S_d	29.00	28.60	.40	-1.602
S_c	28.50	27.50	1.00	0.000

[All values in K ohms affect characteristic but not mantissa. Thus, log scale relationship is unaltered though log scale points are reduced by a constant of 3. For instance, 29.80K = 29,800 and 28.20K = 28,200. Thus, 29,800 − 28,200 = 1,600. And, log 1,600 = 3.204 instead of 0.204.

FIGURE 15.10

In order to compare group observations, mean per cent change and mean log change have been obtained for the observations on four subjects shown in Table 15.4. The two mean distributions are shown superimposed in Figure 15.11.

	BR	MD
	84	77
	81	77
Subject 1	80	76
	78	74
	77	74
	72	63
	73	61
Subject 2	67	60
	64	60
	66	63
	67	59
	65	61
Subject 3	64	60
	58	53
	51	48
	54	45
	52	45
Subject 4	54	48
	52	48
	55	52

TABLE 15.4

Normalized Transformations

In addition to magnitude change, per cent change (resistance or conductance), and log change, it is also possible to transform the distribution of GSR observations by means of other mathematical operations, such as square root, etc. But these operations do not add measurably to the usefulness of the transformed data distributions. If sequential dependency can be ignored for the moment, and a normalizing transformation can be employed, then the distribution should fit a parametric model with greater fidelity. Such a transformation has been described by Lacey (see Bibliography). The resulting normalized scores are called *autonomic lability scores* (ALS).

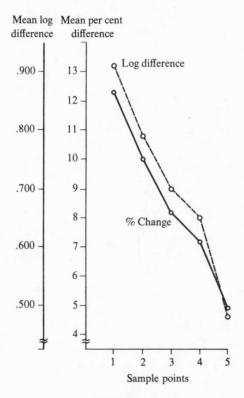

FIGURE 15.11

The computation of the ALS is based on the degree of correlation between the basal resistance (BR) and the *resistance level at maximum deflection* (MD), or between BR and GSR.

We can consider the BR and MD distributions as the X and Y distributions, and compute r_{XY}. The coefficient, r_{XY}, and the ALS for each of the obtained measures is computed in Table 15.5.

BR X	MD Y	x	x^2	y	y^2	xy	Z_X	Z_Y
29.8	28.2	+.525	.28	+.15	.02	+.08	+.95	+.38
29.8	27.9	+.525	.28	−.15	.02	−.08	+.95	−.38
29.0	28.6	−.275	.08	+.55	.30	−.15	−.50	+1.38
28.5	27.5	−.775	.60	−.55	.30	+.43	−1.41	−1.38
117.1	112.2		1.24		.64	+.28		

$$\bar{X} = 29.275$$

$$\bar{Y} = 28.05$$

$$S_X{}^2 = \frac{1.24}{4} = .31 \qquad S_Y{}^2 = \frac{.64}{4} = .16$$

$$S_X = \sqrt{.31} = .55 \qquad S_Y = \sqrt{.16} = .40$$

$$r_{XY} = \frac{.28/4}{.55(.40)} = \frac{.07}{.22} = .32$$

$$ALS = 50 + 10 \left[\frac{Z_Y - Z_X(r_{XY})}{\sqrt{(1 - r_{XY}{}^2)}} \right]$$

$$ALS_1 = 50 + 10 \left[\frac{.38 - .95(.32)}{\sqrt{[1 - .32(.32)]}} \right] = 50 + 10 \left[\frac{.38 - .95(.32)}{\sqrt{(1 - .1)}} \right]$$

$$= 50 + 10 \left(\frac{.38 - .30}{.95} \right)$$

$$= 50 + 10(.084) = 50.84$$

$$ALS_2 = 42.84$$

$$ALS_3 = 66.20$$

$$ALS_4 = 40.21$$

TABLE 15.5

In Figure 15.12, the ALS distribution is shown compared to the per cent conductance change and log resistance change for the same observations. The three respective distributions are plotted so that the midpoint of the range of the transformed observations coincides (approximately) with the middle of the ordinate scales. Units have been expanded so the ordinate scales match, and unit differences do not coincide from scale to scale; the figure simply serves to show the nature of the transformations.

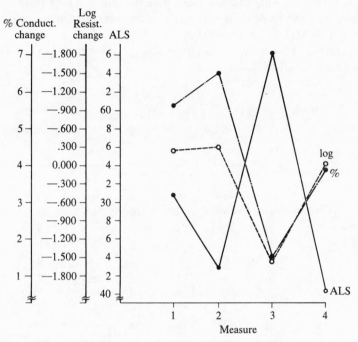

FIGURE 15.12

Measuring Sequential Dependency: λ

Most investigators tend to ignore one property of many empirical observation distributions, even though they are well aware of its importance. This
property is the relationship between successive responses of subjects, i.e.
sequential dependency. By definition, sequential dependency at the least
violates the assumption that each observation is independently obtained from
the population. Also, it is quite likely that the degree of sequential dependency is a major feature of the difference between populations. One consequence of ignoring sequential dependency is that this form of bias distorts
any interpretation of critical ratios in tests of significant differences between
populations studied.

The traditional use of the mean and standard deviation of distributions to
summarize populations of observations ignores any indication of the degree
of sequential dependency. The use of mean and standard deviation is not
related to the nature of response sequence. A statistic sensitive to first-
order sequential dependency is the λ (Greek lambda) statistic, described by
Lathrop (see Bibliography) in connection with psychophysical and signal

detection tasks and with compensatory psychomotor tracking tasks. When applied to the analysis of GSR patterns, it is an index of change in reference points: BR to MD to BR to MD . . ., etc.

The λ statistic is basically a measure of *average absolute* slope of successive response measures, and is defined as

$$\lambda = \sqrt{\frac{\Sigma|X - X_a|}{(n - 1)\sigma}}$$

n	X_a	x_a	x_a^2	X_b	x_b	x_b^2
1	1	-4	16	1	-4	16
2	2	-3	9	9	$+4$	16
3	3	-2	4	2	-3	9
4	4	-1	1	8	$+3$	9
5	5	0	0	3	-2	4
6	6	$+1$	1	7	$+2$	4
7	7	$+2$	4	4	-1	1
8	8	$+3$	9	6	$+1$	1
9	9	$+4$	16	5	0	0

$$\Sigma = 45 \qquad 0 \qquad 60 \quad \Sigma = 45 \qquad 0 \qquad 60$$

$$\bar{X}_a = 5 \qquad\qquad\qquad \bar{X}_b = 5$$

$$S^2 = \frac{\Sigma x^2}{n}$$

$$= \frac{60}{9}$$

$$= 6.67$$

$$S = 2.58$$

$$\lambda_a = \sqrt{\frac{1 - 2}{8(2.58)} + \frac{2 - 3}{8(2.58)} + \frac{3 - 4}{8(2.58)} \cdots + \frac{8 - 9}{8(2.58)}}$$

$$= \sqrt{.384}$$

$$= .62$$

$$\lambda_b = \sqrt{\frac{1 - 9}{8(2.58)} + \frac{9 - 2}{8(2.58)} + \frac{2 - 8}{8(2.58)} + \frac{8 - 3}{8(2.58)} \cdots + \frac{6 - 5}{8(2.58)}}$$

$$= \sqrt{1.74}$$

$$= 1.32$$

TABLE 15.6

where X is any observation, and X_a is the next observation in the sequence. When λ's are computed for n length sequences of random normal numbers, the parameters of the distribution are

$$\mu_\lambda = 1.00 \qquad \text{and} \qquad \sigma_\lambda = \frac{1}{2\sqrt{n}}$$

Distributions with stable and relatively small sequential differences are in the lower tail of the theoretical distribution, and alternating sequences are in the upper tail. The midpoint of the distribution represents those sequences with no significant first-order sequential dependency.

To illustrate the use and computation of λ, the data in Table 15.6 are employed. Note that there are two distributions, X_a and X_b, whose mean and standard deviation are the same. The distributions are plotted in Figure 15.13. These distributions are distributions of consecutive observations.

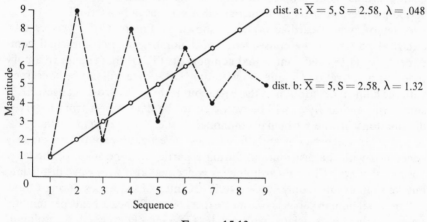

FIGURE 15.13

It is quite clear that the sequences are dramatically different. Yet $\overline{X}$ and S are identical. Thus $\overline{X}$ and S would hardly be adequate if we wished to describe the characteristic property of each distribution—the order in which the observations occur.

The application of this analysis of stability of sequential observations to GSR data is based on the contention that when GSR = 0, BR = MD. The changes in the pattern of this autonomic response measure constitute consecutive BR observations, and resistance readings at any point (arbitrarily designated as BR, the reference point, or as MD, the point of maximum deflection) are simply consecutive though dependently related observations.

The data in Figure 15.1 may be employed to demonstrate the application of λ to GSR analysis.

The sequences in question consists of the consecutive BR — MD, MD — BR readings. For this set of sequences, λ is 1.18. When k sets of such sequences are observed, the within mean square (MS_w), rather than S, is used to describe the distribution of k — λ measures. Distributions of λ may appropriately be compared by means of analysis of variance methods.

The λ coefficient is superior to indices based on average error or root mean square. The latter measures rely on error amplitude primarily, while λ, though it retains an amplitude factor, adds an instability factor.

The Cardiac Response

The recording of the activity of the heart, unlike that made for the analysis of changes in skin resistance, is a recording of current generated by the myoneural cardiac tissue proper. In addition, cardiac activity shows a series of periodic cycles, whereas spontaneous GSR's are usually aperiodic.

A set of typical cardiac cycles is shown in Figure 15.2. This cardiac electrical pattern is the consequence of a number of precontractual components (P, QRS) and contractual components (T, U) that change in amplitude and polarization ($+$ and $-$) with time lapse. We will not be concerned with the clinical description of the EKG, but rather with gross changes, EKG summaries and analyses will be restricted to the data we get from observed distributions of one particular component: the R component of the QRS complex. In the behavior and life sciences, investigators are often primarily concerned with the distribution, during a particular time span, of the large positive R wave. Clinical cardiology is concerned also with wave distortion, but we shall examine only successive R - R interval sequences during t time.

In normal human subjects one may expect 60 to 100 heart-beats per minute. Each heartbeat is associated with one P - QRS - T - U cycle. A recording of cardiac activity lasting 10 seconds is shown in Figure 15.14. The broken line at the beginning of the record indicates the beginning of the time period. The broken line at the end indicates the end of the 10-second span. The typical EKG record is obtained on paper marked in millimeter units. Each heavy line (time line) represents .20 second (the chart drive is 25 mm per second). Seconds elapsed are noted in large numerals at the top of the record.

Standard evaluation of cardiac rate is made in the following manner. A QRS wave complex close to a time line is chosen. (Note the 15 time lines indicated on the recording in Figure 15.14.) The interval between each QRS complex represents a cycle, and fractions of time between QRS complexes represent fractions of a cycle. The number of QRS complexes (plus the fraction, if there should be one) multiplied by 20 is equal to the number of *ventricular* contractions falling within the estimated 60-second period. Multiplication by 20 is used for *beats-per-minute* (BPM) transformation.

In many behavior and life science studies, investigators find the measures of BPM, successive BPM, and the *interbeat time interval* (IBI) most useful. Other valuable indices of change in cardiac cycle patterns are per cent change, transformation to ALS (as in GSR analysis), or to a special ratio.

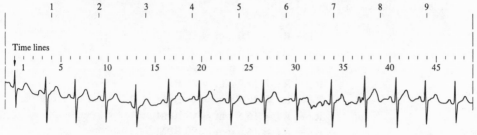

FIGURE 15.14

Estimated Beats per Minute

There are alternatives to counting the number of beats (R waves) in the entire obtained record and dividing by t time. Figure 15.14 shows 10 seconds of EKG record. Note that there are 15 beats during this 10-second span. The BPM can be estimated by multiplying 15 by 6 (6 × 10 seconds = 1 minute). On this basis the BPM is 90. The clinical method is based on 15 time lines, i.e. 15 × .20 second, or 3 seconds. Note that there are $4\frac{1}{2}$ cycles within the 15 time lines, and $4\frac{1}{2}$ multiplied by 20 is 90.

Successive Beats per Minute

Ordinarily, EKG recording are obtained at a standard chart speed of 25 mm/sec. The accuracy of the measurement of the distance between successive BPM is greatly increased when the chart drive speed is greater than 25 mm/sec; however, it can be estimated at the standard speed rate. Successive BPM are determined in this fashion. At 25 mm/sec, 1500 mm/min are recorded. The 1500-mm span, divided by the number of R waves, is the BPM measure for any successive R - R interval point.

It might be desirable to know not only the rate per minute expected for a *hypothetical* minute, but also at what rate per minute the heart is actually beating throughout a *given* minute, since one might expect the rate to vary during that minute. In this case, the successive BPM can be determined in the following way. Take any two successive R waves, and measure the distance between the two in millimeters. Determine what value you need, so that when 1500 is divided by this value the resulting quantity is the distance in mm between the two successive R waves. For instance, in Figure 15.14, the distance between the two first R waves in the record is 17 mm. 1500 divided by 88 is approximately 17. Thus, at the time of the second beat, the heart of the subject would be beating at 88 BPM, were it to continue at that rate. The distance between the second and third R - R waves is 15.5 mm. 1500 divided by 97 is approximately 15.5. Thus, by the third beat, the heart was beating at 97 BPM. The distribution of successive BPM is computed in Table 15.8 and plotted in Figure 15.15.

$$R_1 - R_2 = 17.0 \text{ mm} = 88 \text{ BPM}$$
$$R_2 - R_3 = 15.5 \text{ mm} = 97 \text{ BPM}$$
$$R_3 - R_4 = 16.0 \text{ mm} = 94 \text{ BPM}$$
$$R_4 - R_5 = 17.0 \text{ mm} = 88 \text{ BPM}$$
$$R_5 - R_6 = 17.5 \text{ mm} = 86 \text{ BPM}$$
$$R_6 - R_7 = 16.5 \text{ mm} = 90 \text{ BPM}$$
$$R_7 - R_8 = 16.5 \text{ mm} = 90 \text{ BPM}$$
$$R_8 - R_9 = 17.5 \text{ mm} = 86 \text{ BPM}$$
$$R_9 - R_{10} = 17.5 \text{ mm} = 86 \text{ BPM}$$
$$R_{10} - R_{11} = 18.5 \text{ mm} = 81 \text{ BPM}$$
$$R_{11} - R_{12} = 17.5 \text{ mm} = 86 \text{ BPM}$$
$$R_{12} - R_{13} = 16.5 \text{ mm} = 90 \text{ BPM}$$
$$R_{13} - R_{14} = 16.0 \text{ mm} = 94 \text{ BPM}$$
$$R_{14} - R_{15} = 16.0 \text{ mm} = 94 \text{ BPM}$$

TABLE 15.8

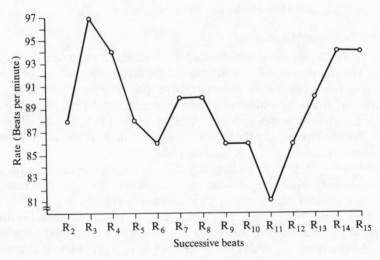

FIGURE 15.15

The mean of the distribution of 15 successive measures of BPM is 83.33. This value is probably closer than that obtained by other estimates. It should be noted that this successive BPM measure is based on the interbeat time interval (IBI). The corresponding IBI's for the above distribution are shown in Figure 15.15.

Per Cent and Relative Per Cent Change

Sample cardiac cycles can be compared by employing change and per cent change. The procedure for obtaining per cent change is similar to that used in obtaining the GSR. For illustrative purposes, successive per cent change of the data in Figure 15.14 are shown in Figure 15.16.

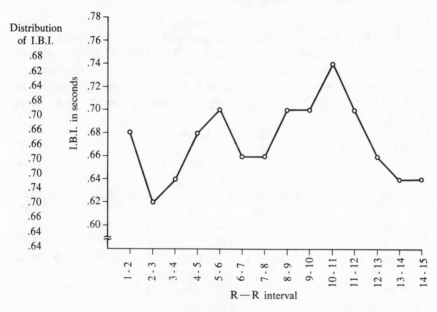

Distribution of I.B.I.
.68
.62
.64
.68
.70
.66
.66
.70
.70
.74
.70
.66
.64
.64

FIGURE 15.16

Another form of per cent change has been suggested for comparing cardiac activity samples. This procedure* results in *per cent change relative to maximum*

$$\frac{C - P}{max - P}$$

where C is the heart rate at a particular point, P is the rate to which C is being compared, and max is the maximum heart rate during an arbitrary reference or basal activity sample. The procedure is appropriate for both

* Described in Black *et al.*; see Bibliography.

within- and between-subjects comparisons. For instance, we might wish to compare the 11th BPM rate to the 1st BPM rate in the data in Figure 15.14. The 11th cycle rate is 81 BPM, the first estimated BPM rate is for the second cycle, and that rate is 88 BPM. The maximum rate in the record is 97 BPM. Thus,

$$\% \text{ change relative to maximum} = \frac{81 - 88}{97 - 88}$$

$$= \frac{7}{11}$$

$$= 65\%$$

Normalized Transformation: ALS

The cardiac ALS* computations are illustrated below, employing the data in Table 15.9. The observations are arranged in two distributions: an arbitrary pre-stress and post-stress group. The BPM rates for the two groups are shown in Figure 15.17(a). A bar-graph (histogram) is used here because these are not successive rates in subjects or subject groups and no trend can be implied. The corresponding ALS measures are shown in Figure 15.17(b).

Heart Rate (BPM)

(X) Pre-stress	(Y) Post-stress	x	y	x^2	y^2	xy	Z_X	Z_Y	ALS
65	110	-13	$+4$	169	16	-52	-1.40	$+.36$	55.86
85	95	$+7$	-11	49	121	-77	$+.76$	$-.99$	38.80
70	105	-8	-1	64	1	$+8$	$-.86$	$-.09$	50.51
90	125	$+12$	$+19$	144	361	$+228$	$+1.29$	$+1.71$	65.20
80	95	$+2$	-11	4	121	-22	$+.22$	$-.99$	39.60
$\Sigma = 390$	530	0	0	430	620	85			

$\bar{X} = 78.0 \quad \bar{Y} = 106.0$

$$S_X{}^2 = \frac{430}{5} = 86; \quad S_X = \sqrt{86} = 9.27$$

$$S_Y{}^2 = \frac{620}{5} = 124; \quad S_Y = \sqrt{124} = 11.1$$

$$r_{XY} = \frac{85/5}{9.27(11.1)} = \frac{17.0}{102.9} = .16$$

$$\text{ALS} = 50 + 10 \left[\frac{Z_Y - Z_X(r_{XY})}{\sqrt{1 - r_{XY}{}^2}} \right] = 50 + 10 \left[\frac{Z_Y - Z_X(.16)}{.99} \right]$$

TABLE 15.9

* See Lacey in the Bibliography.

It should be pointed out that all of the preceding methods resulting in distributions of transformed observations do not indicate, *per se*, that the transformations permit the use of parametric analyses for group comparisons. Tests for trend and trend differences employing critical ratios based on percentiles of the F distributions can be applied only when the distributions do not depart markedly from normality. Investigators have been encouraged, in the past, to make unwarranted assumptions about the degree of departure

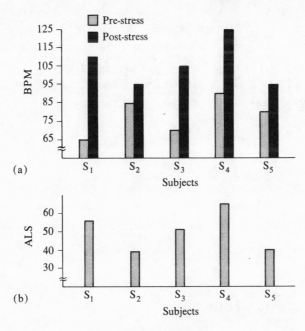

FIGURE 15.17

of certain populations from normality which the statistical tests could tolerate. There are numerous citations in the scientific literature supporting the contention that "moderate" departure of the parent population from the normal distribution form often results only in a slight misstatement of the probability of the critical ratio obtained. There is also, however, much evidence contraindicating the validity of this contention.

Transformations may simplify data summaries and displays. But appropriate statistical comparisons can be made only to transformations that result in distributions meeting the specified criteria for the analysis of variance. Of these transformations, the ALS seems to be the closest.

Problems

Figure 15.18 is a record of the change in resistance (GSR) observed in a human subject, under standard recording conditions, when a 1000 cps tone (stimulus) was presented. Transform the first 10 GSR's, i.e. S_1, S_2, $S_3 \ldots S_{10}$, to (1) per cent change, (2) conductance, (3) log difference, and (4) ALS.

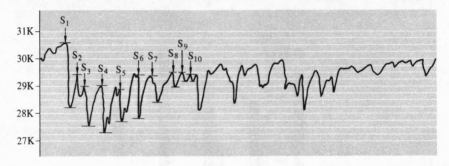

FIGURE 15.18

Answers

1. Transformation to per cent change:

	BR	MD	(BR − MD)	$\left(\dfrac{BR - MD}{BR}\right)$
S_1	30500	28125	2375	.078
S_2	29375	28625	750	.026
S_3	29000	27500	1500	.052
S_4	29000	27250	1750	.060
S_5	28875	27625	1250	.043
S_6	29375	27875	1500	.051
S_7	29375	28375	1000	.034
S_8	29500	29000	500	.017
S_9	29500	29175	325	.011
S_{10}	29375	29125	250	.009

2. Transformation to conductance units:

	$\dfrac{1}{BR}$	$\dfrac{1}{MD}$	$\left(\dfrac{1}{BR} - \dfrac{1}{MD}\right)$
S_1	.0000328	.0000356	.0000028
S_2	.0000340	.0000349	.0000009
S_3	.0000345	.0000364	.0000019
S_4	.0000345	.0000367	.0000022
S_5	.0000346	.0000362	.0000016
S_6	.0000340	.0000359	.0000019
S_7	.0000340	.0000352	.0000012
S_8	.0000339	.0000345	.0000006
S_9	.0000339	.0000343	.0000004
S_{10}	.0000340	.0000343	.0000003

3. Transformation to log difference:

	(BR − MD)	(log BR − MD)
S_1	2375	3.3766
S_2	750	2.8751
S_3	1500	3.1761
S_4	1750	3.2430
S_5	1250	3.0969
S_6	1500	3.1761
S_7	1000	3.0000
S_8	500	2.6990
S_9	325	2.5119
S_{10}	250	2.3979

260   INTRODUCTION TO STATISTICS

4. Transformation to ALS:

X	X²	x	Zx	Y	Y²	y	Zy	xy
30.50	930.25	+1.11	+2.64	28.12	790.73	−0.15	−0.23	−0.17
29.38	863.18	−0.01	−0.02	28.62	819.10	+0.35	+0.54	0
29.00	841.00	−0.39	−0.93	27.50	756.25	−0.77	−1.18	+0.30
29.00	841.00	−0.39	−0.93	27.25	742.56	−1.02	−1.56	+0.40
28.88	834.05	−0.51	−1.21	27.62	762.86	−0.65	0	+0.33
29.38	863.18	−0.01	−0.02	27.88	777.29	−0.39	−0.60	0
29.38	863.18	−0.01	−0.02	28.38	805.42	+0.11	+0.17	0
29.50	870.25	+0.11	+0.26	29.00	841.00	+0.73	+1.12	+0.08
29.50	870.25	+0.11	+0.26	29.18	851.47	+0.91	+1.40	+0.10
29.38	863.18	−0.01	−0.02	29.12	847.97	+0.85	+1.31	0

$\Sigma = 293.90$

$\Sigma = 8639.52$

$\Sigma = 282.67$

$\Sigma = 7994.65$

$\Sigma = +1.04$

$$\bar{X} = \frac{239.90}{10} = 23.39; \qquad S_X = 0.42$$

$$\bar{Y} = \frac{282.67}{10} = 28.27; \qquad S_Y = 0.65$$

$$r_{XY} = \frac{\dfrac{1.04}{10}}{(0.42)(0.65)}$$

$$= \frac{0.10}{0.27}$$

$$= .37$$

$$ALS = 50 + 10 \left[\frac{Z_Y - Z_X(r_{XY})}{\sqrt{[1 - (r_{XY})^2]}} \right]$$

$$= 50 + 10 \left[\frac{Z_Y - Z_X(.37)}{.93} \right]$$

$$ALS_1 = 44.00$$
$$ALS_2 = 53.76$$
$$ALS_3 = 43.23$$
$$ALS_4 = 40.54$$
$$ALS_5 = 45.05$$
$$ALS_6 = 47.42$$
$$ALS_7 = 51.18$$
$$ALS_8 = 57.42$$
$$ALS_9 = 59.35$$
$$ALS_{10} = 59.14$$

Appendix

Table A.1. Normal distribution: ordinate Y at $\pm Z$ and area a between $\pm Z$.

Z	X	Y	a	1 − a	Z	X	Y	a	1 − a
0	μ	.399	.0000	1.0000	±1.50	$\mu \pm 1.50\sigma$	.1295	.8664	.1336
± .05	$\mu \pm .05\sigma$	.398	.0399	.9601	±1.55	$\mu \pm 1.55\sigma$	.1200	.8789	.1211
± .10	$\mu \pm .10\sigma$	.397	.0797	.9203	±1.60	$\mu \pm 1.60\sigma$	.1109	.8904	.1096
± .15	$\mu \pm .15\sigma$	.394	.1192	.8808	±1.65	$\mu \pm 1.65\sigma$	.1023	.9011	.0989
± .20	$\mu \pm .20\sigma$	.391	.1585	.8415	±1.70	$\mu \pm 1.70\sigma$	.0940	.9109	.0891
± .25	$\mu \pm .25\sigma$	.387	.1974	.8026	±1.75	$\mu \pm 1.75\sigma$	.0863	.9199	.0801
± .30	$\mu \pm .30\sigma$	.381	.2358	.7642	±1.80	$\mu \pm 1.80\sigma$	.0790	.9281	.0719
± .35	$\mu \pm .35\sigma$	.375	.2737	.7263	±1.85	$\mu \pm 1.85\sigma$	.0721	.9357	.0643
± .40	$\mu \pm .40\sigma$	.368	.3108	.6892	±1.90	$\mu \pm 1.90\sigma$	.0656	.9426	.0574
± .45	$\mu \pm .45\sigma$	.361	.3473	.6527	±1.95	$\mu \pm 1.95\sigma$	.0596	.9488	.0512
± .50	$\mu \pm .50\sigma$	.352	.3829	.6171	±2.00	$\mu \pm 2.00\sigma$	.0540	.9545	.0455
± .55	$\mu \pm .55\sigma$	.343	.4177	.5823	±2.05	$\mu \pm 2.05\sigma$	.0488	.9596	.0404
± .60	$\mu \pm .60\sigma$	.333	.4515	.5485	±2.10	$\mu \pm 2.10\sigma$	.0440	.9643	.0357
± .65	$\mu \pm .65\sigma$	.323	.4843	.5157	±2.15	$\mu \pm 2.15\sigma$	.0396	.9684	.0316
± .70	$\mu \pm .70\sigma$	.312	.5161	.4839	±2.20	$\mu \pm 2.20\sigma$	.0355	.9722	.0278
± .75	$\mu \pm .75\sigma$	.301	.5467	.4533	±2.25	$\mu \pm 2.25\sigma$	.0317	.9756	.0244
± .80	$\mu \pm .80\sigma$	.290	.5763	.4237	±2.30	$\mu \pm 2.30\sigma$	.0283	.9786	.0214
± .85	$\mu \pm .85\sigma$	.278	.6047	.3953	±2.35	$\mu \pm 2.35\sigma$	.0252	.9812	.0188
± .90	$\mu \pm .90\sigma$	.266	.6319	.3681	±2.40	$\mu \pm 2.40\sigma$	.0224	.9836	.0164
± .95	$\mu \pm .95\sigma$	.254	.6579	.3421	±2.45	$\mu \pm 2.45\sigma$	.0198	.9857	.0143
±1.00	$\mu \pm 1.00\sigma$	.242	.6827	.3173	±2.50	$\mu \pm 2.50\sigma$	.0175	.9876	.0124
±1.05	$\mu \pm 1.05\sigma$	.230	.7063	.2937	±2.55	$\mu \pm 2.55\sigma$	.0154	.9892	.0108
±1.10	$\mu \pm 1.10\sigma$	.218	.7287	.2713	±2.60	$\mu \pm 2.60\sigma$	.0136	.9907	.0093
±1.15	$\mu \pm 1.15\sigma$	.206	.7499	.2501	±2.65	$\mu \pm 2.65\sigma$	.0119	.9920	.0080
±1.20	$\mu \pm 1.20\sigma$	.194	.7699	.2301	±2.70	$\mu \pm 2.70\sigma$	.0104	.9931	.0069
±1.25	$\mu \pm 1.25\sigma$	.183	.7887	.2113	±2.75	$\mu \pm 2.75\sigma$	.0091	.9940	.0060
±1.30	$\mu \pm 1.30\sigma$	.171	.8064	.1936	±2.80	$\mu \pm 2.80\sigma$	.0079	.9949	.0051
±1.35	$\mu \pm 1.35\sigma$	.160	.8230	.1770	±2.85	$\mu \pm 2.85\sigma$	.0069	.9956	.0044
±1.40	$\mu \pm 1.40\sigma$	.150	.8385	.1615	±2.90	$\mu \pm 2.90\sigma$	.0060	.9963	.0037
±1.45	$\mu \pm 1.45\sigma$	.139	.8529	.1471	±2.95	$\mu \pm 2.95\sigma$	.0051	.9968	.0032
±1.50	$\mu \pm 1.50\sigma$	.130	.8664	.1336	±3.00	$\mu \pm 3.00\sigma$	.0044	.9973	.0027
					±4.00	$\mu \pm 4.00\sigma$	.0001	.99994	.00006
					±5.00	$\mu \pm 5.00\sigma$	.000001	.9999994	.0000006
± .000	μ	.3989	.0000	1.0000	±1.036	$\mu \pm 1.036\sigma$	.2331	.7000	.3000
± .126	$\mu \pm .126\sigma$	.3958	.1000	.9000	±1.282	$\mu \pm 1.282\sigma$	.1755	.8000	.2000
± .253	$\mu \pm .253\sigma$	.3863	.2000	.8000	±1.645	$\mu \pm 1.645\sigma$	.1031	.9000	.1000
± .385	$\mu \pm .385\sigma$	.3704	.3000	.7000	±1.960	$\mu \pm 1.960\sigma$	.0584	.9500	.0500
± .524	$\mu \pm .524\sigma$	.3477	.4000	.6000	±2.576	$\mu \pm 2.576\sigma$	.0145	.9900	.0100
± .674	$\mu \pm .674\sigma$	.3178	.5000	.5000	±3.291	$\mu \pm 3.291\sigma$	.0018	.9990	.0010
± .842	$\mu \pm .842\sigma$	.2800	.6000	.4000	±3.891	$\mu \pm 3.891\sigma$	.0002	.9999	.0001

[From W. J. Dixon and F. J. Massey, *Introduction to Statistical Analysis*. Copyright © 1957 by McGraw-Hill, Inc. Used by permission of McGraw-Hill Book Company.]

Table A.2. Cumulative normal distribution.

Z	X	Area	Z	X	Area
−3.25	$\mu - 3.25\sigma$	.0006	−1.00	$\mu - 1.00\sigma$	.1587
−3.20	$\mu - 3.20\sigma$	.0007	− .95	$\mu - .95\sigma$	.1711
−3.15	$\mu - 3.15\sigma$	.0008	− .90	$\mu - .90\sigma$	.1841
−3.10	$\mu - 3.10\sigma$	.0010	− .85	$\mu - .85\sigma$	.1977
−3.05	$\mu - 3.05\sigma$	.0011	− .80	$\mu - .80\sigma$	.2119
−3.00	$\mu - 3.00\sigma$	.0013	− .75	$\mu - .75\sigma$	.2266
−2.95	$\mu - 2.95\sigma$	.0016	− .70	$\mu - .70\sigma$	.2420
−2.90	$\mu - 2.90\sigma$	.0019	− .65	$\mu - .65\sigma$	.2578
−2.85	$\mu - 2.85\sigma$	.0022	− .60	$\mu - .60\sigma$	.2743
−2.80	$\mu - 2.80\sigma$	.0026	− .55	$\mu - .55\sigma$	.2912
−2.75	$\mu - 2.75\sigma$	.0030	− .50	$\mu - .50\sigma$	.3085
−2.70	$\mu - 2.70\sigma$	.0035	− .45	$\mu - .45\sigma$	.3264
−2.65	$\mu - 2.65\sigma$	.0040	− .40	$\mu - .40\sigma$	.3446
−2.60	$\mu - 2.60\sigma$	.0047	− .35	$\mu - .35\sigma$	.3632
−2.55	$\mu - 2.55\sigma$	.0054	− .30	$\mu - .30\sigma$	.3821
−2.50	$\mu - 2.50\sigma$	.0062	− .25	$\mu - .25\sigma$	.4013
−2.45	$\mu - 2.45\sigma$	.0071	− .20	$\mu - .20\sigma$	.4207
−2.40	$\mu - 2.40\sigma$	.0082	− .15	$\mu - .15\sigma$	.4404
−2.35	$\mu - 2.35\sigma$	.0094	− .10	$\mu - .10\sigma$	.4602
−2.30	$\mu - 2.30\sigma$	.0107	− .05	$\mu - .05\sigma$	.4801
−2.25	$\mu - 2.25\sigma$	.0122			
−2.20	$\mu - 2.20\sigma$	.0139			
−2.15	$\mu - 2.15\sigma$	.0158	.00	μ	.5000
−2.10	$\mu - 2.10\sigma$	.0179			
−2.05	$\mu - 2.05\sigma$	.0202			
−2.00	$\mu - 2.00\sigma$	.0228	.05	$\mu + .05\sigma$	.5199
−1.95	$\mu - 1.95\sigma$	.0256	.10	$\mu + .10\sigma$	.5398
−1.90	$\mu - 1.90\sigma$	.0287	.15	$\mu + .15\sigma$	.5596
−1.85	$\mu - 1.85\sigma$	.0322	.20	$\mu + .20\sigma$	.5793
−1.80	$\mu - 1.80\sigma$	.0359	.25	$\mu + .25\sigma$	.5987
−1.75	$\mu - 1.75\sigma$	.0401	.30	$\mu + .30\sigma$	.6179
−1.70	$\mu - 1.70\sigma$	.0446	.35	$\mu + .35\sigma$	.6368
−1.65	$\mu - 1.65\sigma$	.0495	.40	$\mu + .40\sigma$	.6554
−1.60	$\mu - 1.60\sigma$	.0548	.45	$\mu + .45\sigma$	.6736
−1.55	$\mu - 1.55\sigma$	.0606	.50	$\mu + .50\sigma$	.6915
−1.50	$\mu - 1.50\sigma$	.0668	.55	$\mu + .55\sigma$	.7088
−1.45	$\mu - 1.45\sigma$	.0735	.60	$\mu + .60\sigma$	.7257
−1.40	$\mu - 1.40\sigma$	.0808	.65	$\mu + .65\sigma$	.7422
−1.35	$\mu - 1.35\sigma$	.0885	.70	$\mu + .70\sigma$	.7580
−1.30	$\mu - 1.30\sigma$	.0968	.75	$\mu + .75\sigma$	.7734
−1.25	$\mu - 1.25\sigma$	.1056	.80	$\mu + .80\sigma$	.7881
−1.20	$\mu - 1.20\sigma$	.1151	.85	$\mu + .85\sigma$	.8023
−1.15	$\mu - 1.15\sigma$	.1251	.90	$\mu + .90\sigma$	.8159
−1.10	$\mu - 1.10\sigma$	.1357	.95	$\mu + .95\sigma$	.8289
−1.05	$\mu - 1.05\sigma$	.1469	1.00	$\mu + 1.00\sigma$	.8413

[From W. J. Dixon and F. J. Massey, *Introduction to Statistical Analysis*. Copyright ©

Z	X	Area	Z	X	Area
1.05	$\mu + 1.05\sigma$	.8531	−4.265	$\mu - 4.265\sigma$	.00001
1.10	$\mu + 1.10\sigma$	.8643	−3.719	$\mu - 3.719\sigma$	.0001
1.15	$\mu + 1.15\sigma$	.8749	−3.090	$\mu - 3.090\sigma$	.001
1.20	$\mu + 1.20\sigma$	.8849	−2.576	$\mu - 2.576\sigma$	.005
1.25	$\mu + 1.25\sigma$	.8944	−2.326	$\mu - 2.326\sigma$	.01
1.30	$\mu + 1.30\sigma$	.9032	−2.054	$\mu - 2.054\sigma$	.02
1.35	$\mu + 1.35\sigma$	.9115	−1.960	$\mu - 1.960\sigma$	.025
1.40	$\mu + 1.40\sigma$	.9192	−1.881	$\mu - 1.881\sigma$	.03
1.45	$\mu + 1.45\sigma$	.9265	−1.751	$\mu - 1.751\sigma$	.04
1.50	$\mu + 1.50\sigma$	.9332	−1.645	$\mu - 1.645\sigma$	.05
1.55	$\mu + 1.55\sigma$	.9394	−1.555	$\mu - 1.555\sigma$	.06
1.60	$\mu + 1.60\sigma$	.9452	−1.476	$\mu - 1.476\sigma$	.07
1.65	$\mu + 1.65\sigma$	.9505	−1.405	$\mu - 1.405\sigma$	.08
1.70	$\mu + 1.70\sigma$	.9554	−1.341	$\mu - 1.341\sigma$	.09
1.75	$\mu + 1.75\sigma$	.9599	−1.282	$\mu - 1.282\sigma$	.10
1.80	$\mu + 1.80\sigma$	.9641	−1.036	$\mu - 1.036\sigma$	.15
1.85	$\mu + 1.85\sigma$	.9678	− .842	$\mu - .842\sigma$	.20
1.90	$\mu + 1.90\sigma$	.9713	− .674	$\mu - .674\sigma$	.25
1.95	$\mu + 1.95\sigma$	.9744	− .524	$\mu - .524\sigma$	.30
2.00	$\mu + 2.00\sigma$	.9772	− .385	$\mu - .385\sigma$	.35
2.05	$\mu + 2.05\sigma$	.9798	− .253	$\mu - .253\sigma$	.40
2.10	$\mu + 2.10\sigma$	.9821	− .126	$\mu - .126\sigma$	.45
2.15	$\mu + 2.15\sigma$	.9842	0	μ	.50
2.20	$\mu + 2.20\sigma$	.9861	.126	$\mu + .126\sigma$	.55
2.25	$\mu + 2.25\sigma$	.9878	.253	$\mu + .253\sigma$	.60
2.30	$\mu + 2.30\sigma$	.9893	.385	$\mu + .385\sigma$	.65
2.35	$\mu + 2.35\sigma$	.9906	.524	$\mu + .524\sigma$	.70
2.40	$\mu + 2.40\sigma$	.9918	.674	$\mu + .674\sigma$	.75
2.45	$\mu + 2.45\sigma$	.9929	.842	$\mu + .842\sigma$	.80
2.50	$\mu + 2.50\sigma$	.9938	1.036	$\mu + 1.036\sigma$	.85
2.55	$\mu + 2.55\sigma$	.9946	1.282	$\mu + 1.282\sigma$	.90
2.60	$\mu + 2.60\sigma$	.9953	1.341	$\mu + 1.341\sigma$	.91
2.65	$\mu + 2.65\sigma$	.9960	1.405	$\mu + 1.405\sigma$	.92
2.70	$\mu + 2.70\sigma$	.9965	1.476	$\mu + 1.476\sigma$	.93
2.75	$\mu + 2.75\sigma$	.9970	1.555	$\mu + 1.555\sigma$	.94
2.80	$\mu + 2.80\sigma$	.9974	1.645	$\mu + 1.645\sigma$	.95
2.85	$\mu + 2.85\sigma$	.9978	1.751	$\mu + 1.751\sigma$	.96
2.90	$\mu + 2.90\sigma$	.9981	1.881	$\mu + 1.881\sigma$	.97
2.95	$\mu + 2.95\sigma$	.9984	1.960	$\mu + 1.960\sigma$	.975
3.00	$\mu + 3.00\sigma$	.9987	2.054	$\mu + 2.051\sigma$	.98
3.05	$\mu + 3.05\sigma$	.9989	2.326	$\mu + 2.326\sigma$	.99
3.10	$\mu + 3.10\sigma$	.9990	2.576	$\mu + 2.576\sigma$	.995
3.15	$\mu + 3.15\sigma$	.9992	3.090	$\mu + 3.090\sigma$	.999
3.20	$\mu + 3.20\sigma$	.9993	3.719	$\mu + 3.719\sigma$	.9999
3.25	$\mu + 3.25\sigma$	.9994	4.265	$\mu + 4.265\sigma$	.99999

Table A.3. Percentile values of the unit normal curve.

Area to the left of Z	Z	Area to the left of Z	Z	Area to the left of Z	Z	Area to the left of Z	Z	Area to the left of Z	Z
.0001	−3.719	.045	−1.695	.280	−.583	.700	.524	.950	1.645
.0002	−3.540	.050	−1.645	.300	−.524	.720	.583	.955	1.695
.0003	−3.432	.055	−1.598	.320	−.468	.740	.643	.960	1.751
.0004	−3.353	.060	−1.555	.340	−.412	.750	.674	.965	1.812
.0005	−3.291	.065	−1.514	.360	−.358	.760	.706	.970	1.881
.001	−3.000	.070	−1.476	.380	−.305	.780	.772	.975	1.960
.002	−2.878	.075	−1.440	.400	−.253	.800	.842	.980	2.054
.003	−2.748	.080	−1.405	.420	−.202	.820	.915	.985	2.170
.004	−2.652	.085	−1.372	.440	−.151	.840	.994	.990	2.326
.005	−2.576	.090	−1.341	.460	−.100	.860	1.080	.991	2.366
.006	−2.512	.095	−1.311	.480	−.050	.880	1.175	.992	2.409
.007	−2.457	.100	−1.282	.500	.000	.900	1.282	.993	2.457
.008	−2.409	.120	−1.175	.520	.050	.905	1.311	.994	2.512
.009	−2.366	.140	−1.080	.540	.100	.910	1.341	.995	2.576
.010	−2.326	.160	− .994	.560	.151	.915	1.372	.996	2.652
.015	−2.170	.180	− .915	.580	.202	.920	1.405	.997	2.748
.020	−2.054	.200	− .842	.600	.253	.925	1.440	.998	2.878
.025	−1.960	.220	− .772	.620	.305	.930	1.476	.999	3.090
.030	−1.881	.240	− .706	.640	.358	.935	1.514	.9995	3.291
.035	−1.812	.250	− .674	.660	.412	.940	1.555	.9996	3.353
.040	−1.751	.260	− .643	.680	.468	.945	1.598	.9999	3.719

[From Table II of A. L. Bernstein, *A Handbook of Statistics Solutions for the Behavioral Sciences.* Holt, Rinehart and Winston, New York, 1964. Used by permission.]

Table A.4. Percentiles of the t distribution.

df	$t_{.60}$	$t_{.70}$	$t_{.80}$	$t_{.90}$	$t_{.95}$	$t_{.975}$	$t_{.99}$	$t_{.995}$
1	.325	.727	1.376	3.078	6.314	12.706	31.821	63.657
2	.289	.617	1.061	1.886	2.920	4.303	6.965	9.925
3	.277	.584	.978	1.638	2.353	3.182	4.541	5.841
4	.271	.569	.941	1.533	2.132	2.776	3.747	4.604
5	.267	.559	.920	1.476	2.015	2.571	3.365	4.032
6	.265	.553	.906	1.440	1.943	2.447	3.143	3.707
7	.263	.549	.896	1.415	1.895	2.365	2.998	3.499
8	.262	.546	.889	1.397	1.860	2.306	2.896	3.355
9	.261	.543	.883	1.383	1.833	2.262	2.821	3.250
10	.260	.542	.879	1.372	1.812	2.228	2.764	3.169
11	.260	.540	.876	1.363	1.796	2.201	2.718	3.106
12	.259	.539	.873	1.356	1.782	2.179	2.681	3.055
13	.259	.538	.870	1.350	1.771	2.160	2.650	3.012
14	.258	.537	.868	1.345	1.761	2.145	2.624	2.977
15	.258	.536	.866	1.341	1.753	2.131	2.602	2.947
16	.258	.535	.865	1.337	1.746	2.120	2.583	2.921
17	.257	.534	.863	1.333	1.740	2.110	2.567	2.898
18	.257	.534	.862	1.330	1.734	2.101	2.552	2.878
19	.257	.533	.861	1.328	1.729	2.093	2.539	2.861
20	.257	.533	.860	1.325	1.725	2.086	2.528	2.845
21	.257	.532	.859	1.323	1.721	2.080	2.518	2.831
22	.256	.532	.858	1.321	1.717	2.074	2.508	2.819
23	.256	.532	.858	1.319	1.714	2.069	2.500	2.807
24	.256	.531	.857	1.318	1.711	2.064	2.492	2.797
25	.256	.531	.856	1.316	1.708	2.060	2.485	2.787
26	.256	.531	.856	1.315	1.706	2.056	2.479	2.779
27	.256	.531	.855	1.314	1.703	2.052	2.473	2.771
28	.256	.530	.855	1.313	1.701	2.048	2.467	2.763
29	.256	.530	.854	1.311	1.699	2.045	2.462	2.756
30	.256	.530	.854	1.310	1.697	2.042	2.457	2.750
40	.255	.529	.851	1.303	1.684	2.021	2.423	2.704
60	.254	.527	.848	1.296	1.671	2.000	2.390	2.660
120	.254	.526	.845	1.289	1.658	1.980	2.358	2.617
∞	.253	.524	.842	1.282	1.645	1.960	2.326	2.576
df	$-t_{.40}$	$-t_{.30}$	$-t_{.20}$	$-t_{.10}$	$-t_{.05}$	$-t_{.025}$	$-t_{.01}$	$-t_{.005}$

[From Table III of Sir Ronald A. Fisher and Dr. Frank Yates, *Statistical Tables for Biological, Agricultural and Medical Research*, 6th Ed., 1963. Oliver and Boyd Ltd., Edinburgh. By permission of the authors and publishers.]

Table A.5(a). Distribution of χ^2.

df	.99	.98	.95	.90	.80	.70	.50	.30	.20	.10	.05	.02	.01	.001
1	$.0^3157$	$.0^3628$	.00393	.0158	.0642	.148	.455	1.074	1.642	2.706	3.841	5.412	6.635	10.827
2	.0201	.0404	.103	.211	.446	.713	1.386	2.408	3.219	4.605	5.991	7.824	9.210	13.815
3	.115	.185	.352	.584	1.005	1.424	2.366	3.665	4.642	6.251	7.815	9.837	11.345	16.268
4	.297	.429	.711	1.064	1.649	2.195	3.357	4.878	5.989	7.779	9.488	11.668	13.277	18.465
5	.554	.752	1.145	1.610	2.343	3.000	4.351	6.064	7.289	9.236	11.070	13.388	15.086	20.517
6	.872	1.134	1.635	2.204	3.070	3.828	5.348	7.231	8.558	10.645	12.592	15.033	16.812	22.457
7	1.239	1.564	2.167	2.833	3.822	4.671	6.346	8.383	9.803	12.017	14.067	16.622	18.475	24.322
8	1.646	2.032	2.733	3.490	4.594	5.527	7.344	9.524	11.030	13.362	15.507	18.168	20.090	26.125
9	2.088	2.532	3.325	4.168	5.380	6.393	8.343	10.656	12.242	14.684	16.919	19.679	21.666	27.877
10	2.558	3.059	3.940	4.865	6.179	7.267	9.342	11.781	13.442	15.987	18.307	21.161	23.209	29.588
11	3.053	3.609	4.575	5.578	6.989	8.148	10.341	12.899	14.631	17.275	19.675	22.618	24.725	31.264
12	3.571	4.178	5.226	6.304	7.807	9.034	11.340	14.011	15.812	18.549	21.026	24.054	26.217	32.909
13	4.107	4.765	5.892	7.042	8.634	9.926	12.340	15.119	16.985	19.812	22.362	25.472	27.688	34.528
14	4.660	5.368	6.571	7.790	9.467	10.821	13.339	16.222	18.151	21.064	23.685	26.873	29.141	36.123
15	5.229	5.985	7.261	8.547	10.307	11.721	14.339	17.322	19.311	22.307	24.996	28.259	30.578	37.697
16	5.812	6.614	7.962	9.312	11.152	12.624	15.338	18.418	20.465	23.542	26.296	29.633	32.000	39.252
17	6.408	7.255	8.672	10.085	12.002	13.531	16.338	19.511	21.615	24.769	27.587	30.995	33.409	40.790
18	7.015	7.906	9.390	10.865	12.857	14.440	17.338	20.601	22.760	25.989	28.869	32.346	34.805	42.312
19	7.633	8.567	10.117	11.651	13.716	15.352	18.338	21.689	23.900	27.204	30.144	33.687	36.191	43.820
20	8.260	9.237	10.851	12.443	14.578	16.266	19.337	22.775	25.038	28.412	31.410	35.020	37.566	45.315
21	8.897	9.915	11.591	13.240	15.445	17.182	20.337	23.858	26.171	29.615	32.671	36.343	38.932	46.797
22	9.542	10.600	12.338	14.041	16.314	18.101	21.337	24.939	27.301	30.813	33.924	37.659	40.289	48.268
23	10.196	11.293	13.091	14.848	17.187	19.021	22.337	26.018	28.429	32.007	35.172	38.968	41.638	49.728
24	10.856	11.992	13.848	15.659	18.062	19.943	23.337	27.096	29.553	33.196	36.415	40.270	42.980	51.179
25	11.524	12.697	14.611	16.473	18.940	20.867	24.337	28.172	30.675	34.382	37.652	41.566	44.314	52.620
26	12.198	13.409	15.379	17.292	19.820	21.792	25.336	29.246	31.795	35.563	38.885	42.856	45.642	54.052
27	12.879	14.125	16.151	18.114	20.703	22.719	26.336	30.319	32.912	36.741	40.113	44.140	46.963	55.476
28	13.565	14.847	16.928	18.939	21.588	23.647	27.336	31.391	34.027	37.916	41.337	45.419	48.278	56.893
29	14.256	15.574	17.708	19.768	22.475	24.577	28.336	32.461	35.139	39.087	42.557	46.693	49.588	58.302
30	14.953	16.306	18.493	20.599	23.364	25.508	29.336	33.530	36.250	40.256	43.773	47.962	50.892	59.703

Probability

[From Table IV of Sir Ronald A. Fisher and Dr. Frank Yates, *Statistical Tables for Biological, Agricultural and Medical Research*, 6th Ed., 1963. Oliver and Boyd Ltd., Edinburgh. By permission of the authors and publishers.]

Table A.5(b). Percentiles of the χ^2/df Distributions

df	1.0	2.5	5.0	10	90	95	97.5	99	df
1	$.0^316$	$.0^398$	$.0^239$	.016	2.71	3.84	5.02	6.64	1
2	.010	.025	.052	.106	2.30	3.00	3.69	4.61	2
3	.038	.072	.117	.195	2.08	2.60	3.12	3.78	3
4	.074	.121	.178	.266	1.94	2.37	2.79	3.32	4
5	.111	.166	.229	.322	1.85	2.21	2.57	3.02	5
6	.145	.206	.272	.367	1.77	2.10	2.41	2.80	6
7	.177	.241	.310	.405	1.72	2.01	2.29	2.64	7
8	.206	.272	.342	.436	1.67	1.94	2.19	2.51	8
9	.232	.300	.369	.463	1.63	1.88	2.11	2.41	9
10	.256	.325	.394	.487	1.60	1.83	2.05	2.32	10
11	.278	.347	.416	.507	1.57	1.79	1.99	2.25	11
12	.298	.367	.436	.525	1.55	1.75	1.94	2.18	12
13	.316	.385	.453	.542	1.52	1.72	1.90	2.13	13
14	.333	.402	.469	.556	1.50	1.69	1.87	2.08	14
15	.349	.418	.484	.570	1.49	1.67	1.83	2.04	15
16	.363	.432	.498	.582	1.47	1.64	1.80	2.00	16
17	.377	.445	.510	.593	1.46	1.62	1.78	1.97	17
18	.390	.457	.522	.604	1.44	1.60	1.75	1.93	18
19	.402	.469	.532	.613	1.43	1.59	1.73	1.90	19
20	.413	.480	.543	.622	1.42	1.57	1.71	1.88	20
22	.434	.499	.561	.638	1.40	1.54	1.67	1.83	22
24	.452	.517	.577	.652	1.38	1.52	1.64	1.79	24
26	.469	.532	.592	.665	1.37	1.50	1.61	1.76	26
26	.469	.532	.592	.665	1.37	1.50	1.61	1.76	26
28	.484	.547	.605	.676	1.35	1.48	1.59	1.72	28
30	.498	.560	.616	.687	1.34	1.46	1.57	1.70	30
35	.529	.588	.642	.708	1.32	1.42	1.52	1.64	35
40	.554	.611	.663	.726	1.30	1.39	1.48	1.59	40
45	.576	.630	.680	.741	1.28	1.37	1.45	1.55	45
50	.594	.647	.695	.754	1.26	1.35	1.43	1.52	50
55	.610	.662	.708	.765	1.25	1.33	1.41	1.50	55
60	.625	.675	.720	.774	1.24	1.32	1.39	1.47	60
70	.649	.697	.739	.790	1.22	1.29	1.36	1.43	70
80	.669	.714	.755	.803	1.21	1.27	1.33	1.40	80
90	.686	.729	.768	.814	1.20	1.26	1.31	1.38	90
100	.701	.742	.779	.824	1.18	1.24	1.30	1.36	100
120	.724	.763	.798	.839	1.17	1.22	1.27	1.32	120
140	.743	.780	.812	.850	1.16	1.20	1.25	1.30	140
160	.758	.793	.824	.860	1.15	1.19	1.23	1.28	160
180	.771	.804	.833	.868	1.14	1.18	1.22	1.26	180
200	.782	.814	.841	.874	1.13	1.17	1.21	1.25	200
250	.804	.832	.858	.887	1.12	1.15	1.18	1.22	250
300	.820	.846	.870	.897	1.11	1.14	1.17	1.20	300
350	.833	.857	.879	.904	1.10	1.13	1.15	1.18	350
400	.843	.866	.887	.911	1.09	1.12	1.14	1.17	400
450	.852	.874	.893	.916	1.09	1.11	1.13	1.16	450
500	.859	.880	.898	.920	1.08	1.11	1.13	1.15	500
750	.884	.901	.917	.934	1.07	1.09	1.10	1.12	750
1000	.899	.914	.928	.943	1.06	1.07	1.09	1.11	1000
5000	.954	.961	.967	.974	1.02	1.03	1.04	1.05	5000
∞	1	1	1	1	1	1	1	1	∞

[Adapted from Table A−6b of W. J. Dixon and F. J. Massey, *Introduction to Statistical Analysis,* McGraw-Hill Book Co., Copyright © 1957. Used by permission of McGraw-Hill Book Co.]

Table A.6. Critical values for Cochran's test for homogeneity of variances.

df for s^2	$1-\alpha$	Number of variances										
		2	3	4	5	6	7	8	9	10	15	20
1	.95	.9985	.9669	.9065	.8412	.7808	.7271	.6798	.6385	.6020	.4709	.3894
	.99	.9999	.9933	.9676	.9279	.8828	.8376	.7945	.7544	.7175	.5747	.4799
2	.95	.9750	.8709	.7679	.6838	.6161	.5612	.5157	.4775	.4450	.3346	.2705
	.99	.9950	.9423	.8643	.7885	.7218	.6644	.6152	.5727	.5358	.4069	.3297
3	.95	.9392	.7977	.6841	.5981	.5321	.4800	.4377	.4027	.3733	.2758	.2205
	.99	.9794	.8831	.7814	.6957	.6258	.5685	.5209	.4810	.4469	.3317	.2654
4	.95	.9057	.7457	.6287	.5441	.4803	.4307	.3910	.3584	.3311	.2419	.1921
	.99	.9586	.8335	.7212	.6329	.5635	.5080	.4627	.4251	.3934	.2882	.2288
5	.95	.8772	.7071	.5895	.5065	.4447	.3974	.3595	.3286	.3029	.2195	.1735
	.99	.9373	.7933	.6761	.5875	.5195	.4659	.4226	.3870	.3572	.2593	.2048
6	.95	.8534	.6771	.5598	.4783	.4184	.3726	.3362	.3067	.2823	.2034	.1602
	.99	.9172	.7606	.6410	.5531	.4866	.4347	.3932	.3592	.3308	.2386	.1877
7	.95	.8332	.6530	.5365	.4564	.3980	.3535	.3185	.2901	.2666	.1911	.1501
	.99	.8988	.7335	.6129	.5259	.4608	.4105	.3704	.3378	.3106	.2228	.1748
8	.95	.8159	.6333	.5175	.4387	.3817	.3384	.3043	.2768	.2541	.1815	.1422
	.99	.8823	.7107	.5897	.5037	.4401	.3911	.3522	.3207	.2945	.2104	.1646
9	.95	.8010	.6167	.5017	.4241	.3682	.3259	.2926	.2659	.2439	.1736	.1357
	.99	.8674	.6912	.5702	.4854	.4229	.3751	.3373	.3067	.2813	.2002	.1567
16	.95	.7341	.5466	.4366	.3645	.3135	.2756	.2462	.2226	.2032	.1429	.1108
	.99	.7949	.6059	.4884	.4094	.3529	.3105	.2779	.2514	.2297	.1612	.1248
36	.95	.6602	.4748	.3720	.3066	.2612	.2278	.2022	.1820	.1655	.1144	.0879
	.99	.7067	.5153	.4057	.3351	.2858	.2494	.2214	.1992	.1811	.1251	.0960
144	.95	.5813	.4031	.3093	.2513	.2119	.1833	.1616	.1446	.1308	.0889	.0675
	.99	.6062	.4230	.3251	.2644	.2229	.1929	.1700	.1521	.1376	.0934	.0709

[From Table A.7 of J. L. Myers, *Fundamentals of Experimental Design*. Allyn and Bacon, Boston, 1966. Used by permission.]

Table A.7. Per cent points in the distribution of F.

df₂	df₁	1	2	3	4	5	6	8	12	24	∞
1	1 %	4052	4999	5403	5625	5764	5859	5981	6106	6234	6366
	2.5%	647.79	799.50	864.16	899.58	921.85	937.11	956.66	976.71	997.25	1018.30
	5 %	161.45	199.50	215.71	224.58	230.16	233.99	238.88	243.91	249.05	254.32
	10 %	39.86	49.50	53.59	55.83	57.24	58.20	59.44	60.70	62.00	63.33
2	1	98.49	99.00	99.17	99.25	99.30	99.33	99.36	99.42	99.46	99.50
	2.5	38.51	39.00	39.17	39.25	39.30	39.33	39.37	39.42	39.46	39.50
	5	18.51	19.00	19.16	19.25	19.30	19.33	19.37	19.41	19.45	19.50
	10	8.53	9.00	9.16	9.24	9.29	9.33	9.37	9.41	9.45	9.49
3	1	34.12	30.81	29.46	28.71	28.24	27.91	27.49	27.05	26.60	26.12
	2.5	17.44	16.04	15.44	15.10	14.89	14.74	14.54	14.34	14.12	13.90
	5	10.13	9.55	9.28	9.12	9.01	8.94	8.84	8.74	8.64	8.53
	10	5.54	5.46	5.39	5.34	5.31	5.28	5.25	5.22	5.18	5.13
4	1	21.20	18.00	16.69	15.98	15.52	15.21	14.80	14.37	13.93	13.46
	2.5	12.22	10.65	9.98	9.60	9.36	9.20	8.98	8.75	8.51	8.26
	5	7.71	6.94	6.59	6.39	6.26	6.16	6.04	5.91	5.77	5.63
	10	4.54	4.32	4.19	4.11	4.05	4.01	3.95	3.90	3.83	3.76
5	1	16.26	13.27	12.06	11.39	10.97	10.67	10.29	9.89	9.47	9.02
	2.5	10.01	8.43	7.76	7.39	7.15	6.98	6.76	6.52	6.28	6.02
	5	6.61	5.79	5.41	5.19	5.05	4.95	4.82	4.68	4.53	4.36
	10	4.06	3.78	3.62	3.52	3.45	3.40	3.34	3.27	3.19	3.10
6	1	13.74	10.92	9.78	9.15	8.75	8.47	8.10	7.72	7.31	6.88
	2.5	8.81	7.26	6.60	6.23	5.99	5.82	5.60	5.37	5.12	4.85
	5	5.99	5.14	4.76	4.53	4.39	4.28	4.15	4.00	3.84	3.67
	10	3.78	3.46	3.29	3.18	3.11	3.05	2.98	2.90	2.82	2.72
7	1	12.25	9.55	8.45	7.85	7.46	7.19	6.84	6.47	6.07	5.65
	2.5	8.07	6.54	5.89	5.52	5.29	5.12	4.90	4.67	4.42	4.14
	5	5.59	4.74	4.35	4.12	3.97	3.87	3.73	3.57	3.41	3.23
	10	3.59	3.26	3.07	2.96	2.88	2.83	2.75	2.67	2.58	2.47
8	1	11.26	8.65	7.59	7.01	6.63	6.37	6.03	5.67	5.28	4.86
	2.5	7.57	6.06	5.42	5.05	4.82	4.65	4.43	4.20	3.95	3.67
	5	5.32	4.46	4.07	3.84	3.69	3.58	3.44	3.28	3.12	2.93
	10	3.46	3.11	2.92	2.81	2.73	2.67	2.59	2.50	2.40	2.29

[From Table V of Sir Ronald A. Fisher and Dr. Frank Yates, *Statistical Tables for Biological, Agricultural and Medical Research*, 6th Ed., 1963. Oliver and Boyd Ltd., Edinburgh. By permission of the authors and publishers.]

df$_2$ \ df$_1$		1	2	3	4	5	6	8	12	24	∞
9	1 %	10.56	8.02	6.99	6.42	6.06	5.80	5.47	5.11	4.73	4.31
	2.5%	7.21	5.71	5.08	4.72	4.48	4.32	4.10	3.87	3.61	3.33
	5 %	5.12	4.26	3.86	3.63	3.48	3.37	3.23	3.07	2.90	2.71
	10 %	3.36	3.01	2.81	2.69	2.61	2.55	2.47	2.38	2.28	2.16
10	1	10.04	7.56	6.55	5.99	5.64	5.39	5.06	4.71	4.33	3.91
	2.5	6.94	5.46	4.83	4.47	4.24	4.07	3.85	3.62	3.37	3.08
	5	4.96	4.10	3.71	3.48	3.33	3.22	3.07	2.91	2.74	2.54
	10	3.28	2.92	2.73	2.61	2.52	2.46	2.38	2.28	2.18	2.06
11	1	9.65	7.20	6.22	5.67	5.32	5.07	4.74	4.40	4.02	3.60
	2.5	6.72	5.26	4.63	4.28	4.04	3.88	3.66	3.43	3.17	2.88
	5	4.84	3.98	3.59	3.36	3.20	3.09	2.95	2.79	2.61	2.40
	10	3.23	2.86	2.66	2.54	2.45	2.39	2.30	2.21	2.10	1.97
12	1	9.33	6.93	5.95	5.41	5.06	4.82	4.50	4.16	3.78	3.36
	2.5	6.55	5.10	4.47	4.12	3.89	3.73	3.51	3.28	3.02	2.72
	5	4.75	3.88	3.49	3.26	3.11	3.00	2.85	2.69	2.50	2.30
	10	3.18	2.81	2.61	2.48	2.39	2.33	2.24	2.15	2.04	1.90
13	1	9.07	6.70	5.74	5.20	4.86	4.62	4.30	3.96	3.59	3.16
	2.5	6.41	4.97	4.35	4.00	3.77	3.60	3.39	3.15	2.89	2.60
	5	4.67	3.80	3.41	3.18	3.02	2.92	2.77	2.60	2.42	2.21
	10	3.14	2.76	2.56	2.43	2.35	2.28	2.20	2.10	1.98	1.85
14	1	8.86	6.51	5.56	5.03	4.69	4.46	4.14	3.80	3.43	3.00
	2.5	6.30	4.86	4.24	3.89	3.66	3.50	3.29	3.05	2.79	2.49
	5	4.60	3.74	3.34	3.11	2.96	2.85	2.70	2.53	2.35	2.13
	10	3.10	2.73	2.52	2.39	2.31	2.24	2.15	2.05	1.94	1.80
15	1	8.68	6.36	5.42	4.89	4.56	4.32	4.00	3.67	3.29	2.87
	2.5	6.20	4.77	4.15	3.80	3.58	3.41	3.20	2.96	2.70	2.40
	5	4.54	3.68	3.29	3.06	2.90	2.79	2.64	2.48	2.29	2.07
	10	3.07	2.70	2.49	2.36	2.27	2.21	2.12	2.02	1.90	1.76
16	1	8.53	6.23	5.29	4.77	4.44	4.20	3.89	3.55	3.18	2.75
	2.5	6.12	4.69	4.08	3.73	3.50	3.34	3.12	2.89	2.63	2.32
	5	4.49	3.63	3.24	3.01	2.85	2.74	2.59	2.42	2.24	2.01
	10	3.05	2.67	2.46	2.33	2.24	2.18	2.09	1.99	1.87	1.72
17	1	8.40	6.11	5.18	4.67	4.34	4.10	3.79	3.45	3.08	2.65
	2.5	6.04	4.62	4.01	3.66	3.44	3.28	3.06	2.82	2.56	2.25
	5	4.45	3.59	3.20	2.96	2.81	2.70	2.55	2.38	2.19	1.96
	10	3.03	2.64	2.44	2.31	2.22	2.15	2.06	1.96	1.84	1.69

df₂ \ df₁		1	2	3	4	5	6	8	12	24	∞
18	1 %	8.28	6.01	5.09	4.58	4.25	4.01	3.71	3.37	3.00	2.57
	2.5%	5.98	4.56	3.95	3.61	3.38	3.22	3.01	2.77	2.50	2.19
	5 %	4.41	3.55	3.16	2.93	2.77	2.66	2.51	2.34	2.15	1.92
	10 %	3.01	2.62	2.42	2.29	2.20	2.13	2.04	1.93	1.81	1.66
19	1	8.18	5.93	5.01	4.50	4.17	3.94	3.63	3.30	2.92	2.49
	2.5	5.92	4.51	3.90	3.56	3.33	3.17	2.96	2.72	2.45	2.13
	5	4.38	3.52	3.13	2.90	2.74	2.63	2.48	2.31	2.11	1.88
	10	2.99	2.61	2.40	2.27	2.18	2.11	2.02	1.91	1.79	1.63
20	1	8.10	5.85	4.94	4.43	4.10	3.87	3.56	3.23	2.86	2.42
	2.5	5.87	4.46	3.86	3.51	3.29	3.13	2.91	2.68	2.41	2.09
	5	4.35	3.49	3.10	2.87	2.71	2.60	2.45	2.28	2.08	1.84
	10	2.97	2.59	2.38	2.25	2.16	2.09	2.00	1.89	1.77	1.61
21	1	8.02	5.78	4.87	4.37	4.04	3.81	3.51	3.17	2.80	2.36
	2.5	5.83	4.42	3.82	3.48	3.25	3.09	2.87	2.64	2.37	2.04
	5	4.32	3.47	3.07	2.84	2.68	2.57	2.42	2.25	2.05	1.81
	10	2.96	2.57	2.36	2.23	2.14	2.08	1.98	1.88	1.75	1.59
22	1	7.94	5.72	4.82	4.31	3.99	3.76	3.45	3.12	2.75	2.31
	2.5	5.79	4.38	3.78	3.44	3.22	3.05	2.84	2.60	2.33	2.00
	5	4.30	3.44	3.05	2.82	2.66	2.55	2.40	2.23	2.03	1.78
	10	2.95	2.56	2.35	2.22	2.13	2.06	1.97	1.86	1.73	1.57
23	1	7.88	5.66	4.76	4.26	3.94	3.71	3.41	3.07	2.70	2.26
	2.5	5.75	4.35	3.75	3.41	3.18	3.02	2.81	2.57	2.30	1.97
	5	4.28	3.42	3.03	2.80	2.64	2.53	2.38	2.20	2.00	1.76
	10	2.94	2.55	2.34	2.21	2.11	2.05	1.95	1.84	1.72	1.55
24	1	7.82	5.61	4.72	4.22	3.90	3.67	3.36	3.03	2.66	2.21
	2.5	5.72	4.32	3.72	3.38	3.15	2.99	2.78	2.54	2.27	1.94
	5	4.26	3.40	3.01	2.78	2.62	2.51	2.36	2.18	1.98	1.73
	10	2.93	2.54	2.33	2.19	2.10	2.04	1.94	1.83	1.70	1.53
25	1	7.77	5.57	4.68	4.18	3.86	3.63	3.32	2.99	2.62	2.17
	2.5	5.69	4.29	3.69	3.35	3.13	2.97	2.75	2.51	2.24	1.91
	5	4.24	3.38	2.99	2.76	2.60	2.49	2.34	2.16	1.96	1.71
	10	2.92	2.53	2.32	2.18	2.09	2.02	1.93	1.82	1.69	1.52
26	1	7.72	5.53	4.64	4.14	3.82	3.59	3.29	2.96	2.58	2.13
	2.5	5.66	4.27	3.67	3.33	3.10	2.94	2.73	2.49	2.22	1.88
	5	4.22	3.37	2.98	2.74	2.59	2.47	2.32	2.15	1.95	1.69
	10	2.91	2.52	2.31	2.17	2.08	2.01	1.92	1.81	1.68	1.50

df₂ \ df₁		1	2	3	4	5	6	8	12	24	∞
27	1 %	7.68	5.49	4.60	4.11	3.78	3.56	3.26	2.93	2.55	2.10
	2.5%	5.63	4.24	3.65	3.31	3.08	2.92	2.71	2.47	2.19	1.85
	5 %	4.21	3.35	2.96	2.73	2.57	2.46	2.30	2.13	1.93	1.67
	10 %	2.90	2.51	2.30	2.17	2.07	2.00	1.91	1.80	1.67	1.49
28	1	7.64	5.45	4.57	4.07	3.75	3.53	3.23	2.90	2.52	2.06
	2.5	5.61	4.22	3.63	3.29	3.06	2.90	2.69	2.45	2.17	1.83
	5	4.20	3.34	2.95	2.71	2.56	2.44	2.29	2.12	1.91	1.65
	10	2.89	2.50	2.29	2.16	2.06	2.00	1.90	1.79	1.66	1.48
29	1	7.60	5.42	4.54	4.04	3.73	3.50	3.20	2.87	2.49	2.03
	2.5	5.59	4.20	3.61	3.27	3.04	2.88	2.67	2.43	2.15	1.81
	5	4.18	3.33	2.93	2.70	2.54	2.43	2.28	2.10	1.90	1.64
	10	2.89	2.50	2.28	2.15	2.06	1.99	1.89	1.78	1.65	1.47
30	1	7.56	5.39	4.51	4.02	3.70	3.47	3.17	2.84	2.47	2.01
	2.5	5.57	4.18	3.59	3.25	3.03	2.87	2.65	2.41	2.14	1.79
	5	4.17	3.32	2.92	2.69	2.53	2.42	2.27	2.09	1.89	1.62
	10	2.88	2.49	2.28	2.14	2.05	1.98	1.88	1.77	1.64	1.46
40	1	7.31	5.18	4.31	3.83	3.51	3.29	2.99	2.66	2.29	1.80
	2.5	5.42	4.05	3.46	3.13	2.90	2.74	2.53	2.29	2.01	1.64
	5	4.08	3.23	2.84	2.61	2.45	2.34	2.18	2.00	1.79	1.51
	10	2.84	2.44	2.23	2.09	2.00	1.93	1.83	1.71	1.57	1.38
60	1	7.08	4.98	4.13	3.65	3.34	3.12	2.82	2.50	2.12	1.60
	2.5	5.29	3.93	3.34	3.01	2.79	2.63	2.41	2.17	1.88	1.48
	5	4.00	3.15	2.76	2.52	2.37	2.25	2.10	1.92	1.70	1.39
	10	2.79	2.39	2.18	2.04	1.95	1.87	1.77	1.66	1.51	1.29
120	1	6.85	4.79	3.95	3.48	3.17	2.96	2.66	2.34	1.95	1.38
	2.5	5.15	3.80	3.23	2.89	2.67	2.52	2.30	2.05	1.76	1.31
	5	3.92	3.07	2.68	2.45	2.29	2.17	2.02	1.83	1.61	1.25
	10	2.75	2.35	2.13	1.99	1.90	1.82	1.72	1.60	1.45	1.19
∞	1	6.64	4.60	3.78	3.32	3.02	2.80	2.51	2.18	1.79	1.00
	2.5	5.02	3.69	3.12	2.79	2.57	2.41	2.19	1.94	1.64	1.00
	5	3.84	2.99	2.60	2.37	2.21	2.09	1.94	1.75	1.52	1.00
	10	2.71	2.30	2.08	1.94	1.85	1.77	1.67	1.55	1.38	1.00

Table A.8. Critical values of the Pearson r.

df = n −2*	Level of significance for one-tailed test			
	.05	.025	.01	.005
	Level of significance for two-tailed test			
	.10	.05	.02	.01
1	.9877	.9969	.9995	.9999
2	.9000	.9500	.9800	.9900
3	.8054	.8783	.9343	.9587
4	.7293	.8114	.8822	.9172
5	.6694	.7545	.8329	.8745
6	.6215	.7067	.7887	.8343
7	.5822	.6664	.7498	.7977
8	.5494	.6319	.7155	.7646
9	.5214	.6021	.6851	.7348
10	.4973	.5760	.6581	.7079
11	.4762	.5529	.6339	.6835
12	.4575	.5324	.6120	.6614
13	.4409	.5139	.5923	.6411
14	.4259	.4973	.5742	.6226
15	.4124	.4821	.5577	.6055
16	.4000	.4683	.5425	.5897
17	.3887	.4555	.5285	.5751
18	.3783	.4438	.5155	.5614
19	.3687	.4329	.5034	.5487
20	.3598	.4227	.4921	.5368
25	.3233	.3809	.4451	.4869
30	.2960	.3494	.4093	.4487
35	.2746	.3246	.3810	.4182
40	.2573	.3044	.3578	.3932
45	.2428	.2875	.3384	.3721
50	.2306	.2732	.3218	.3541
60	.2108	.2500	.2948	.3248
70	.1954	.2319	.2737	.3017
80	.1829	.2172	.2565	.2830
90	.1726	.2050	.2422	.2673
100	.1638	.1946	.2301	.2540

* n = number of pairs

Table A.9. Critical values of ρ (rank-order correlation coefficient).

	Level of significance for one-tailed test			
	.05	.025	.01	.005
	Level of significance for two-tailed test			
n*	.10	.05	.02	.01
5	.900	1.000	1.000	—
6	.829	.886	.943	1.000
7	.714	.786	.893	.929
8	.643	.738	.833	.881
9	.600	.683	.783	.833
10	.564	.648	.746	.794
12	.506	.591	.712	.777
14	.456	.544	.645	.715
16	.425	.506	.601	.665
18	.399	.475	.564	.625
20	.377	.450	.534	.591
22	.359	.428	.508	.562
24	.343	.409	.485	.537
26	.329	.392	.465	.515
28	.317	.377	.448	.496
30	.306	.364	.432	.478

* n = number of pairs

[From Table F of R. P. Runyon and A. Haber, *Fundamentals of Behavioral Statistics*. Addison-Wesley, Boston, 1967. Used by permission.]

Table A.10. Values of p/Y for various values of p (normal distribution unit area and unit standard deviation).

p	p/Y	p	p/Y	p	p/Y	p	p/Y	p	p/Y
.01	.3752	.21	.7287	.41	1.0547	.61	1.5899	.81	2.9849
.02	.4131	.22	.7430	.42	1.0745	.62	1.6283	.82	3.1250
.03	.4409	.23	.7575	.43	1.0948	.63	1.6686	.83	3.2799
.04	.4642	.24	.7720	.44	1.1156	.64	1.7107	.84	3.4524
.05	.4848	.25	.7867	.45	1.1369	.65	1.7549	.85	3.6456
.06	.5037	.26	.8016	.46	1.1589	.66	1.8013	.86	3.8638
.07	.5213	.27	.8166	.47	1.1815	.67	1.8501	.87	4.1126
.08	.5382	.28	.8318	.48	1.2047	.68	1.9015	.88	4.3991
.09	.5542	.29	.8472	.49	1.2286	.69	1.9558	.89	4.7331
.10	.5698	.30	.8628	.50	1.2533	.70	2.0133	.90	5.1283
.11	.5850	.31	.8787	.51	1.2788	.71	2.0742	.91	5.6038
.12	.5999	.32	.8948	.52	1.3051	.72	2.1389	.92	6.1884
.13	.6145	.33	.9112	.53	1.3323	.73	2.2078	.93	6.9264
.14	.6290	.34	.9279	.54	1.3604	.74	2.2814	.94	7.8910
.15	.6433	.35	.9449	.55	1.3896	.75	2.3601	.95	9.2111
.16	.6576	.36	.9623	.56	1.4198	.76	2.4447	.96	11.1403
.17	.6718	.37	.9800	.57	1.4512	.77	2.5358	.97	14.2559
.18	.6860	.38	.9980	.58	1.4838	.78	2.6343	.98	20.2404
.19	.7002	.39	1.0165	.59	1.5177	.79	2.7411	.99	37.1454
.20	.7144	.40	1.0353	.60	1.5530	.80	2.8575		

[From Table 16 of E. F. Lindquist, *Statistical Analysis in Educational Research*, Houghton Mifflin, New York, 1940. Used by permission of the publisher.]

Table A.11. Five per cent (lightface) and one per cent (boldface) points of the distribution of W.

			n		
n_j	3	4	5	6	7
3			.689	.645	.615
			.811	**.764**	**.727**
4		.591	.540	.505	.480
		.737	**.669**	**.621**	**.587**
5		.485	.442	.413	.392
		.626	**.563**	**.520**	**.488**
6		.410	.373	.349	.331
		.541	**.484**	**.445**	**.417**
8	.362	.313	.285	.266	.252
	.506	**.424**	**.376**	**.345**	**.323**
10	.292	.250	.230	.214	.203
	.416	**.347**	**.307**	**.281**	**.263**
15	.196	.170	.155	.145	.137
	.288	**.239**	**.211**	**.192**	**.179**
20	.148	.128	.117	.109	.103
	.219	**.181**	**.160**	**.146**	**.136**

[From Table XIV (p. 514) of Allen E. Edwards, *Statistical Methods for the Behavioral Sciences*, Rinehart & Co., New York, 1955. Used by permission.]

Table A.12. Critical values of U and U' for a one-tailed test at $\alpha = 0.05$ or a two-tailed test at $\alpha = 0.10$.

n_2 \ n_1	1	2	3	4	5	6	7	8	9	10	11	12	13	14	15	16	17	18	19	20
1	—	—	—	—	—	—	—	—	—	—	—	—	—	—	—	—	—	—	0 / 19	0 / 20
2	—	—	—	—	0 / 10	0 / 12	0 / 14	1 / 15	1 / 17	1 / 19	1 / 21	2 / 22	2 / 24	2 / 26	3 / 27	3 / 29	3 / 31	4 / 32	4 / 34	4 / 36
3	—	—	0 / 9	0 / 12	1 / 14	2 / 16	2 / 19	3 / 21	3 / 24	4 / 26	5 / 28	5 / 31	6 / 33	7 / 35	7 / 38	8 / 40	9 / 42	9 / 45	10 / 47	11 / 49
4	—	—	0 / 12	1 / 15	2 / 18	3 / 21	4 / 24	5 / 27	6 / 30	7 / 33	8 / 36	9 / 39	10 / 42	11 / 45	12 / 48	14 / 50	15 / 53	16 / 56	17 / 59	18 / 62
5	—	0 / 10	1 / 14	2 / 18	4 / 21	5 / 25	6 / 29	8 / 32	9 / 36	11 / 39	12 / 43	13 / 47	15 / 50	16 / 54	18 / 57	19 / 61	20 / 65	22 / 68	23 / 72	25 / 75
6	—	0 / 12	2 / 16	3 / 21	5 / 25	7 / 29	8 / 34	10 / 38	12 / 42	14 / 46	16 / 50	17 / 55	19 / 59	21 / 63	23 / 67	25 / 71	26 / 76	28 / 80	30 / 84	32 / 88
7	—	0 / 14	2 / 19	4 / 24	6 / 29	8 / 34	11 / 38	13 / 43	15 / 48	17 / 53	19 / 58	21 / 63	24 / 67	26 / 72	28 / 77	30 / 82	33 / 86	35 / 91	37 / 96	39 / 101
8	—	1 / 15	3 / 21	5 / 27	8 / 32	10 / 38	13 / 43	15 / 49	18 / 54	20 / 60	23 / 65	26 / 70	28 / 76	31 / 81	33 / 87	36 / 92	39 / 97	41 / 103	44 / 108	47 / 113
9	—	1 / 17	3 / 24	6 / 30	9 / 36	12 / 42	15 / 48	18 / 54	21 / 60	24 / 66	27 / 72	30 / 78	33 / 84	36 / 90	39 / 96	42 / 102	45 / 108	48 / 114	51 / 120	54 / 126
10	—	1 / 19	4 / 26	7 / 33	11 / 39	14 / 46	17 / 53	20 / 60	24 / 66	27 / 73	31 / 79	34 / 86	37 / 93	41 / 99	44 / 106	48 / 112	51 / 119	55 / 125	58 / 132	62 / 138
11	—	1 / 21	5 / 28	8 / 36	12 / 43	16 / 50	19 / 58	23 / 65	27 / 72	31 / 79	34 / 87	38 / 94	42 / 101	46 / 108	50 / 115	54 / 122	57 / 130	61 / 137	65 / 144	69 / 151
12	—	2 / 22	5 / 31	9 / 39	13 / 47	17 / 55	21 / 63	26 / 70	30 / 78	34 / 86	38 / 94	42 / 102	47 / 109	51 / 117	55 / 125	60 / 132	64 / 140	68 / 148	72 / 156	77 / 163
13	—	2 / 24	6 / 33	10 / 42	15 / 50	19 / 59	24 / 67	28 / 76	33 / 84	37 / 93	42 / 101	47 / 109	51 / 118	56 / 126	61 / 134	65 / 143	70 / 151	75 / 159	80 / 167	84 / 176
14	—	2 / 26	7 / 35	11 / 45	16 / 54	21 / 63	26 / 72	31 / 81	36 / 90	41 / 99	46 / 108	51 / 117	56 / 126	61 / 135	66 / 144	71 / 153	77 / 161	82 / 170	87 / 179	92 / 188
15	—	3 / 27	7 / 38	12 / 48	18 / 57	23 / 67	28 / 77	33 / 87	39 / 96	44 / 106	50 / 115	55 / 125	61 / 134	66 / 144	72 / 153	77 / 163	83 / 172	88 / 182	94 / 191	100 / 200
16	—	3 / 29	8 / 40	14 / 50	19 / 61	25 / 71	30 / 82	36 / 92	42 / 102	48 / 112	54 / 122	60 / 132	65 / 143	71 / 153	77 / 163	83 / 173	89 / 183	95 / 193	101 / 203	107 / 213
17	—	3 / 31	9 / 42	15 / 53	20 / 65	26 / 76	33 / 86	39 / 97	45 / 108	51 / 119	57 / 130	64 / 140	70 / 151	77 / 161	83 / 172	89 / 183	96 / 193	102 / 204	109 / 214	115 / 225
18	—	4 / 32	9 / 45	16 / 56	22 / 68	28 / 80	35 / 91	41 / 103	48 / 114	55 / 123	61 / 137	68 / 148	75 / 159	82 / 170	88 / 182	95 / 193	102 / 204	109 / 215	116 / 226	123 / 237
19	0 / 19	4 / 34	10 / 47	17 / 59	23 / 72	30 / 84	37 / 96	44 / 108	51 / 120	58 / 132	65 / 144	72 / 156	80 / 167	87 / 179	94 / 191	101 / 203	109 / 214	116 / 226	123 / 238	130 / 250
20	0 / 20	4 / 36	11 / 49	18 / 62	25 / 75	32 / 88	39 / 101	47 / 113	54 / 126	62 / 138	69 / 151	77 / 163	84 / 176	92 / 188	100 / 200	107 / 213	115 / 225	123 / 237	130 / 250	138 / 262

[From H. B. Mann and D. R. Whitney, *Ann. Math. Statist.*, 18, 52-54, 1947. Used by permission.]

Table A.13. Estimates of r_{tet} for various values of ad/bc.

r_{tet}	ad/bc	r_{tet}	ad/bc	r_{tet}	ad/bc
.00	0–1.00	.35	2.49–2.55	.70	8.50–8.90
.01	1.01–1.03	.36	2.56–2.63	.71	8.91–9.35
.02	1.04–1.06	.37	2.64–2.71	.72	9.36–9.82
.03	1.07–1.08	.38	2.72–2.79	.73	9.83–10.33
.04	1.09–1.11	.39	2.80–2.87	.74	10.34–10.90
.05	1.12–1.14	.40	2.88–2.96	.75	10.91–11.51
.06	1.15–1.17	.41	2.97–3.05	.76	11.52–12.16
.07	1.18–1.20	.42	3.06–3.14	.77	12.17–12.89
.08	1.21–1.23	.43	3.15–3.24	.78	12.90–13.70
.09	1.24–1.27	.44	3.25–3.34	.79	13.71–14.58
.10	1.28–1.30	.45	3.35–3.45	.80	14.59–15.57
.11	1.31–1.33	.46	3.46–3.56	.81	15.58–16.65
.12	1.34–1.37	.47	3.57–3.68	.82	16.66–17.88
.13	1.38–1.40	.48	3.69–3.80	.83	17.89–19.28
.14	1.41–1.44	.49	3.81–3.92	.84	19.29–20.85
.15	1.45–1.48	.50	3.93–4.06	.85	20.86–22.68
.16	1.49–1.52	.51	4.07–4.20	.86	22.69–24.76
.17	1.53–1.56	.52	4.21–4.34	.87	24.77–27.22
.18	1.57–1.60	.53	4.35–4.49	.88	27.23–30.09
.19	1.61–1.64	.54	4.50–4.66	.89	30.10–33.60
.20	1.65–1.69	.55	4.67–4.82	.90	33.61–37.79
.21	1.70–1.73	.56	4.83–4.99	.91	37.80–43.06
.22	1.74–1.78	.57	5.00–5.18	.92	43.07–49.83
.23	1.79–1.83	.58	5.19–5.38	.93	49.84–58.79
.24	1.84–1.88	.59	5.39–5.59	.94	58.80–70.95
.25	1.89–1.93	.60	5.60–5.80	.95	70.96–89.01
.26	1.94–1.98	.61	5.81–6.03	.96	89.02–117.54
.27	1.99–2.04	.62	6.04–6.28	.97	117.55–169.67
.28	2.05–2.10	.63	6.29–6.54	.98	169.68–293.12
.29	2.11–2.15	.64	6.55–6.81	.99	293.13–923.97
.30	2.16–2.22	.65	6.82–7.10	1.00	923.98 . . .
.31	2.23–2.28	.66	7.11–7.42		
.32	2.29–2.34	.67	7.43–7.75		
.33	2.35–2.41	.68	7.76–8.11		
.34	2.42–2.48	.69	8.12–8.49		

[From Table VIII of N. M. Downie and R. W. Heath, *Basic Standard Methods*, Harper & Row, New York, 1965. This table was originally computed by M. D. Davidoff and H. W. Goheen, *Psychometrika*, 1953, 18, 115-21. The reader should also consult the "Note," *Psychometrika*, June 1954. The Table here is used by permission of the authors and *Psychometrika*.]

N	1%	5%	10%	25%	N	1%	5%	10%	25%
1					46	13	15	16	18
2					47	14	16	17	19
3				0	48	14	16	17	19
4				0	49	15	17	18	19
5			0	0	50	15	17	18	20
6		0	0	1	51	15	18	19	20
7		0	0	1	52	16	18	19	21
8	0	0	1	1	53	16	18	20	21
9	0	1	1	2	54	17	19	20	22
10	0	1	1	2	55	17	19	20	22
11	0	1	2	3	56	17	20	21	23
12	1	2	2	3	57	18	20	21	23
13	1	2	3	3	58	18	21	22	24
14	1	2	3	4	59	19	21	22	24
15	2	3	3	4	60	19	21	23	25
16	2	3	4	5	61	20	22	23	25
17	2	4	4	5	62	20	22	24	25
18	3	4	5	6	63	20	23	24	26
19	3	4	5	6	64	21	23	24	26
20	3	5	5	6	65	21	24	25	27
21	4	5	6	7	66	22	24	25	27
22	4	5	6	7	67	22	25	26	28
23	4	6	7	8	68	22	25	26	28
24	5	6	7	8	69	23	25	27	29
25	5	7	7	9	70	23	26	27	29
26	6	7	8	9	71	24	26	28	30
27	6	7	8	10	72	24	27	28	30
28	6	8	9	10	73	25	27	28	31
29	7	8	9	10	74	25	28	29	31
30	7	9	10	11	75	25	28	29	32
31	7	9	10	11	76	26	28	30	32
32	8	9	10	12	77	26	29	30	32
33	8	10	11	12	78	27	29	31	33
34	9	10	11	13	79	27	30	31	33
35	9	11	12	13	80	28	30	32	34
36	9	11	12	14	81	28	31	32	34
37	10	12	13	14	82	28	31	33	35
38	10	12	13	14	83	29	32	33	35
39	11	12	13	15	84	29	32	33	36
40	11	13	14	15	85	30	32	34	36
41	11	13	14	16	86	30	33	34	37
42	12	14	15	16	87	31	33	35	37
43	12	14	15	17	88	31	34	35	38
44	13	15	16	17	89	31	34	36	38
45	13	15	16	18	90	32	35	36	39

Table A.15. Squares and square roots of numbers from 1 to 1000.

Number	Square	Square root	Number	Square	Square root
1	1	1.000	51	26 01	7.141
2	4	1.414	52	27 04	7.211
3	9	1.732	53	28 09	7.280
4	16	2.000	54	29 16	7.348
5	25	2.236	55	30 25	7.416
6	36	2.449	56	31 36	7.483
7	49	2.646	57	32 49	7.550
8	64	2.828	58	33 64	7.616
9	81	3.000	59	34 81	7.681
10	1 00	3.162	60	36 00	7.746
11	1 21	3.317	61	37 21	7.810
12	1 44	3.464	62	38 44	7.874
13	1 69	3.606	63	39 69	7.937
14	1 96	3.742	64	40 96	8.000
15	2 25	3.873	65	42 25	8.062
16	2 56	4.000	66	43 56	8.124
17	2 89	4.123	67	44 89	8.185
18	3 24	4.243	68	46 24	8.246
19	3 61	4.359	69	47 61	8.307
20	4 00	4.472	70	49 00	8.367
21	4 41	4.583	71	50 41	8.426
22	4 84	4.690	72	51 84	8.485
23	5 29	4.796	73	53 29	8.544
24	5 76	4.899	74	54 76	8.602
25	6 25	5.000	75	56 25	8.660
26	6 76	5.099	76	57 76	8.718
27	7 29	5.196	77	59 29	8.775
28	7 84	5.292	78	60 84	8.832
29	8 41	5.385	79	62 41	8.888
30	9 00	5.477	80	64 00	8.944
31	9 61	5.568	81	65 61	9.000
32	10 24	5.657	82	67 24	9.055
33	10 89	5.745	83	68 89	9.110
34	11 56	5.831	84	70 56	9.165
35	12 25	5.916	85	72 25	9.220
36	12 96	6.000	86	73 96	9.274
37	13 69	6.083	87	75 69	9.327
38	14 44	6.164	88	77 44	9.381
39	15 21	6.245	89	79 21	9.434
40	16 00	6.325	90	81 00	9.487
41	16 81	6.403	91	82 81	9.539
42	17 64	6.481	92	84 64	9.592
43	18 49	6.557	93	86 49	9.644
44	19 36	6.633	94	88 36	9.695
45	20 25	6.708	95	90 25	9.747
46	21 16	6.782	96	92 16	9.798
47	22 09	6.856	97	94 09	9.849
48	23 04	6.928	98	96 04	9.899
49	24 01	7.000	99	98 01	9.950
50	25 00	7.071	100	1 00 00	10.000

Number	Square	Square root	Number	Square	Square root
101	1 02 01	10.050	151	2 28 01	12.288
102	1 04 04	10.100	152	2 31 04	12.329
103	1 06 09	10.149	153	2 34 09	12.369
104	1 08 16	10.198	154	2 37 16	12.410
105	1 10 25	10.247	155	2 40 25	12.450
106	1 12 36	10.296	156	2 43 36	12.490
107	1 14 49	10.344	157	2 46 49	12.530
108	1 16 64	10.392	158	2 49 64	12.570
109	1 18 81	10.440	159	2 52 81	12.610
110	1 21 00	10.488	160	2 56 00	12.649
111	1 23 21	10.536	161	2 59 21	12.689
112	1 25 44	10.583	162	2 62 44	12.728
113	1 27 69	10.630	163	2 65 69	12.767
114	1 29 96	10.677	164	2 68 96	12.806
115	1 32 25	10.724	165	2 72 25	12.845
116	1 34 56	10.770	166	2 75 56	12.884
117	1 36 89	10.817	167	2 78 89	12.923
118	1 39 24	10.863	168	2 82 24	12.961
119	1 41 61	10.909	169	2 85 61	13.000
120	1 44 00	10.954	170	2 89 00	13.038
121	1 46 41	11.000	171	2 92 41	13.077
122	1 48 84	11.045	172	2 95 84	13.115
123	1 51 29	11.091	173	2 99 29	13.153
124	1 53 76	11.136	174	3 02 76	13.191
125	1 56 25	11.180	175	3 06 25	13.229
126	1 58 76	11.225	176	3 09 76	13.266
127	1 61 29	11.269	177	3 13 29	13.304
128	1 63 84	11.314	178	3 16 84	13.342
129	1 66 41	11.358	179	3 20 41	13.379
130	1 69 00	11.402	180	3 24 00	13.416
131	1 71 61	11.446	181	3 27 61	13.454
132	1 74 24	11.489	182	3 31 24	13.491
133	1 76 89	11.533	183	3 34 89	13.528
134	1 79 56	11.576	184	3 38 56	13.565
135	1 82 25	11.619	185	3 42 25	13.601
136	1 84 96	11.662	186	3 45 96	13.638
137	1 87 69	11.705	187	3 49 69	13.675
138	1 90 44	11.747	188	3 53 44	13.711
139	1 93 21	11.790	189	3 57 21	13.748
140	1 96 00	11.832	190	3 61 00	13.784
141	1 98 81	11.874	191	3 64 81	13.820
142	2 01 64	11.916	192	3 68 64	13.856
143	2 04 49	11.958	193	3 72 49	13.892
144	2 07 36	12.000	194	3 76 36	13.928
145	2 10 25	12.042	195	3 80 25	13.964
146	2 13 16	12.083	196	3 84 16	14.000
147	2 16 09	12.124	197	3 88 09	14.036
148	2 19 04	12.166	198	3 92 04	14.071
149	2 22 01	12.207	199	3 96 01	14.107
150	2 25 00	12.247	200	4 00 00	14.142

Number	Square	Square root	Number	Square	Square root
201	4 04 01	14.177	251	6 30 01	15.843
202	4 08 04	14.213	252	6 35 04	15.875
203	4 12 09	14.248	253	6 40 09	15.906
204	4 16 16	14.283	254	6 45 16	15.937
205	4 20 25	14.318	255	6 50 25	15.969
206	4 24 36	14.353	256	6 55 36	16.000
207	4 28 49	14.387	257	6 60 49	16.031
208	4 32 64	14.422	258	6 65 64	16.062
209	4 36 81	14.457	259	6 70 81	16.093
210	4 41 00	14.491	260	6 76 00	16.125
211	4 45 21	14.526	261	6 81 21	16.155
212	4 49 44	14.560	262	6 86 44	16.186
213	4 53 69	14.595	263	6 91 69	16.217
214	4 57 96	14.629	264	6 96 96	16.248
215	4 62 25	14.663	265	7 02 25	16.279
216	4 66 56	14.697	266	7 07 56	16.310
217	4 70 89	14.731	267	7 12 89	16.340
218	4 75 24	14.765	268	7 18 24	16.371
219	4 79 61	14.799	269	7 23 61	16.401
220	4 84 00	14.832	270	7 29 00	16.432
221	4 88 41	14.866	271	7 34 41	16.462
222	4 92 84	14.900	272	7 39 84	16.492
223	4 97 29	14.933	273	7 45 29	16.523
224	5 01 76	14.967	274	7 50 76	16.553
225	5 06 25	15.000	275	7 56 25	16.583
226	5 10 76	15.033	276	7 61 76	16.613
227	5 15 29	15.067	277	7 67 29	16.643
228	5 19 84	15.100	278	7 72 84	16.673
229	5 24 41	15.133	279	7 78 41	16.703
230	5 29 00	15.166	280	7 84 00	16.733
231	5 33 61	15.199	281	7 89 61	16.763
232	5 38 24	15.232	282	7 95 24	16.793
233	5 42 89	15.264	283	8 00 89	16.823
234	5 47 56	15.297	284	8 06 56	16.852
235	5 52 25	15.330	285	8 12 25	16.882
236	5 56 96	15.362	286	8 17 96	16.912
237	5 61 69	15.395	287	8 23 69	16.941
238	5 66 44	15.427	288	8 29 44	16.971
239	5 71 21	15.460	289	8 35 21	17.000
240	5 76 00	15.492	290	8 41 00	17.029
241	5 80 81	15.524	291	8 46 81	17.059
242	5 85 64	15.556	292	8 52 64	17.088
243	5 90 49	15.588	293	8 58 49	17.117
244	5 95 36	15.620	294	8 64 36	17.146
245	6 00 25	15.652	295	8 70 25	17.176
246	6 05 16	15.684	296	8 76 16	17.205
247	6 10 09	15.716	297	8 82 09	17.234
248	6 15 04	15.748	298	8 88 04	17.263
249	6 20 01	15.780	299	8 94 10	17.292
250	6 25 00	15.811	300	9 00 00	17.321

Number	Square	Square root	Number	Square	Square root
301	9 06 01	17.349	351	12 32 01	18.735
302	9 12 04	17.378	352	12 39 04	18.762
303	9 18 09	17.407	353	12 46 09	18.788
304	9 24 16	17.436	354	12 53 16	18.815
305	9 30 25	17.464	355	12 60 25	18.841
306	9 36 36	17.493	356	12 67 36	18.868
307	9 42 49	17.521	357	12 74 49	18.894
308	9 48 64	17.550	358	12 81 64	18.921
309	9 54 81	17.578	359	12 88 81	18.947
310	9 61 00	17.607	360	12 96 00	18.974
311	9 67 21	17.635	361	13 03 21	19.000
312	9 73 44	17.664	362	13 10 44	19.026
313	9 79 69	17.692	363	13 17 69	19.053
314	9 85 96	17.720	364	13 24 96	19.079
315	9 92 25	17.748	365	13 32 25	19.105
316	9 98 56	17.776	366	13 39 56	19.131
317	10 04 89	17.804	367	13 46 89	19.157
318	10 11 24	17.833	368	13 54 24	19.183
319	10 17 61	17.861	369	13 61 61	19.209
320	10 24 00	17.889	370	13 69 00	19.235
321	10 30 41	17.916	371	13 76 41	19.261
322	10 36 84	17.944	372	13 83 84	19.287
323	10 43 29	17.972	373	13 91 29	19.313
324	10 49 76	18.000	374	13 98 76	19.339
325	10 56 25	18.028	375	14 06 25	19.363
326	10 62 76	18.055	376	14 13 76	19.391
327	10 69 29	18.083	377	14 21 29	19.416
328	10 75 84	18.111	378	14 28 84	19.442
329	10 82 41	18.138	379	14 36 41	19.468
330	10 89 00	18.166	380	14 44 00	19.494
331	10 95 61	18.193	381	14 51 61	19.519
332	11 02 24	18.221	382	14 59 24	19.545
333	11 08 89	18.248	383	14 66 89	19.570
334	11 15 56	18.276	384	14 74 56	19.596
335	11 22 25	18.303	385	14 82 25	19.621
336	11 28 96	18.330	386	14 89 96	19.647
337	11 35 69	18.358	387	14 97 69	19.672
338	11 42 44	18.385	388	15 05 44	19.698
339	11 49 21	18.412	389	15 13 21	19.723
340	11 56 00	18.439	390	15 21 00	19.748
341	11 62 81	18.466	391	15 28 81	19.774
342	11 69 64	18.493	392	15 36 64	19.799
343	11 76 49	18.520	393	15 44 49	19.824
344	11 83 36	18.547	394	15 52 36	19.849
345	11 90 25	18.574	395	15 60 25	19.875
346	11 97 16	18.601	396	15 68 16	19.900
347	12 04 09	18.628	397	15 76 09	19.925
348	12 11 04	18.655	398	15 84 04	19.950
349	12 18 01	18.682	399	15 92 01	19.975
350	12 25 00	18.708	400	16 00 00	20.000

Number	Square	Square root	Number	Square	Square root
401	16 08 01	20.025	451	20 34 01	21.237
402	16 16 04	20.050	452	20 43 04	21.260
403	16 24 09	20.075	453	20 52 09	21.284
404	16 32 16	20.100	454	20 61 16	21.307
405	16 40 25	20.125	455	20 70 25	21.331
406	16 48 36	20.149	456	20 79 36	21.354
407	16 56 49	20.174	457	20 88 49	21.378
408	16 64 64	20.199	458	20 97 64	21.401
409	16 72 81	20.224	459	21 06 81	21.424
410	16 81 00	20.248	460	21 16 00	21.448
411	16 89 21	20.273	461	21 25 21	21.471
412	16 97 44	20.298	462	21 34 44	21.494
413	17 05 69	20.322	463	21 43 69	21.517
414	17 13 96	20.347	464	21 52 96	21.541
415	17 22 25	20.372	465	21 62 25	21.564
416	17 30 56	20.396	466	21 71 56	21.587
417	17 38 89	20.421	467	21 80 89	21.610
418	17 47 24	20.445	468	21 90 24	21.633
419	17 55 61	20.469	469	21 99 61	21.656
420	17 64 00	20.494	470	22 09 00	21.679
421	17 72 41	20.518	471	22 18 41	21.703
422	17 80 84	20.543	472	22 27 84	21.726
423	17 89 29	20.567	473	22 37 29	21.749
424	17 97 76	20.591	474	22 46 76	21.772
425	18 06 25	20.616	475	22 56 25	21.794
426	18 14 76	20.640	476	22 65 76	21.817
427	18 23 29	20.664	477	22 75 29	21.840
428	18 31 84	20.688	478	22 84 84	21.863
429	18 40 41	20.712	479	22 94 41	21.886
430	18 49 00	20.736	480	23 04 00	21.909
431	18 57 61	20.761	481	23 13 61	21.932
432	18 66 24	20.785	482	23 23 24	21.954
433	18 74 89	20.809	483	23 32 89	21.977
434	18 83 56	20.833	484	23 42 56	22.000
435	18 92 25	20.857	485	23 52 25	22.023
436	19 00 96	20.881	486	23 61 96	22.045
437	19 09 69	20.905	487	23 71 69	22.068
438	19 18 44	20.928	488	23 81 44	22.091
439	19 27 21	20.952	489	23 91 21	22.113
440	19 36 00	20.976	490	24 01 00	22.136
441	19 44 81	21.000	491	24 10 81	22.159
442	19 53 64	21.024	492	24 20 64	22.181
443	19 62 49	21.048	493	24 30 49	22.204
444	19 71 36	21.071	494	24 40 36	22.226
445	19 80 25	21.095	495	24 50 25	22.249
446	19 89 16	21.119	496	24 60 16	22.271
447	19 98 09	21.142	497	24 70 09	22.293
448	20 07 04	21.166	498	24 80 04	22.316
449	20 16 01	21.190	499	24 90 01	22.338
450	20 25 00	21.213	500	25 00 00	22.361

Number	Square	Square root	Number	Square	Square root
501	25 10 01	22.383	551	30 36 01	23.473
502	25 20 04	22.405	552	30 47 04	23.495
503	25 30 09	22.428	553	30 58 09	23.516
504	25 40 16	22.450	554	30 69 16	23.537
505	25 50 25	22.472	555	30 80 25	23.558
506	25 60 36	22.494	556	30 91 36	23.580
507	25 70 49	22.517	557	31 02 49	23.601
508	25 80 64	22.539	558	31 13 64	23.622
509	25 90 81	22.561	559	31 24 81	23.643
510	26 01 00	22.583	560	31 36 00	23.664
511	26 11 21	22.605	561	31 47 21	23.685
512	26 21 44	22.627	562	31 58 44	23.707
513	26 31 69	22.650	563	31 69 69	23.728
514	26 41 96	22.672	564	31 80 96	23.749
515	26 52 25	22.694	565	31 92 25	23.770
516	26 62 56	22.716	566	32 03 56	23.791
517	26 72 89	22.738	567	32 14 89	23.812
518	26 83 24	22.760	568	32 26 24	23.833
519	26 93 61	22.782	569	32 37 61	23.854
520	27 04 00	22.804	570	32 49 00	23.875
521	27 14 41	22.825	571	32 60 41	23.896
522	27 24 84	22.847	572	32 71 84	23.917
523	27 35 29	22.869	573	32 83 29	23.937
524	27 45 76	22.891	574	32 94 76	23.958
525	27 56 25	22.913	575	33 06 25	23.979
526	27 66 76	22.935	576	33 17 76	24.000
527	27 77 29	22.956	577	33 29 29	24.021
528	27 87 84	22.978	578	33 40 84	24.042
529	27 98 41	23.000	579	33 52 41	24.062
530	28 09 00	23.022	580	33 64 00	24.083
531	28 19 61	23.043	581	33 75 61	24.104
532	28 30 24	23.065	582	33 87 24	24.125
533	28 40 89	23.087	583	33 98 89	24.145
534	28 51 56	23.108	584	34 10 56	24.166
535	28 62 25	23.130	585	34 22 25	24.187
536	28 72 96	23.152	586	34 33 96	24.207
537	28 83 69	23.173	587	34 45 69	24.228
538	28 94 44	23.195	588	34 57 44	24.249
539	29 05 21	23.216	589	34 69 21	24.269
540	29 16 00	23.238	590	34 81 00	24.290
541	29 26 81	23.259	591	34 92 81	24.310
542	29 37 64	23.281	592	35 04 64	24.331
543	29 48 49	23.302	593	35 16 49	24.352
544	29 59 36	23.324	594	35 28 36	24.372
545	29 70 25	23.345	595	35 40 25	24.393
546	29 81 16	23.367	596	35 52 16	24.413
547	29 92 09	23.388	597	35 64 09	24.434
548	30 03 04	23.409	598	35 76 04	24.454
549	30 14 01	23.431	599	35 88 01	24.474
550	30 25 00	23.452	600	36 00 00	24.495

Number	Square	Square root	Number	Square	Square root
601	36 12 01	24.515	651	42 38 01	25.515
602	36 24 04	24.536	652	42 51 04	25.534
603	36 36 09	24.556	653	42 64 09	25.554
604	36 48 16	24.576	654	42 77 16	25.573
605	36 60 25	24.597	655	42 90 25	25.593
606	36 72 36	24.617	656	43 03 36	25.612
607	36 84 49	24.637	657	43 16 49	25.632
608	36 96 64	24.658	658	43 29 64	25.652
609	37 08 81	24.678	659	43 42 81	25.671
610	37 21 00	24.698	660	43 56 00	25.690
611	37 33 21	24.718	661	43 69 21	25.710
612	37 45 44	24.739	662	43 82 44	25.729
613	37 57 69	24.759	663	43 95 69	25.749
614	37 69 96	24.779	664	44 08 96	25.768
615	37 82 25	24.799	665	44 22 25	25.788
616	37 94 56	24.819	666	44 35 56	25.807
617	38 06 89	24.839	667	44 48 89	25.826
618	38 19 24	24.860	668	44 62 24	25.846
619	38 31 61	24.880	669	44 75 61	25.865
620	38 44 00	24.900	670	44 89 00	25.884
621	38 56 41	24.920	671	45 02 41	25.904
622	38 68 84	24.940	672	45 15 84	25.923
623	38 81 29	24.960	673	45 29 29	25.942
624	38 93 76	24.980	674	45 42 76	25.962
625	39 06 25	25.000	675	45 56 25	25.981
626	39 18 76	25.020	676	45 69 76	26.000
627	39 31 29	25.040	677	45 83 29	26.019
628	39 43 84	25.060	678	45 96 84	26.038
629	39 56 41	25.080	679	46 10 41	26.058
630	39 69 00	25.100	680	46 24 00	26.077
631	39 81 61	25.120	681	46 37 61	26.096
632	39 94 24	25.140	682	46 51 24	26.115
633	40 06 89	25.159	683	46 64 89	26.134
634	40 19 56	25.179	684	46 78 56	26.153
635	40 32 25	25.199	685	46 92 25	26.173
636	40 44 96	25.219	686	47 05 96	26.192
637	40 57 69	25.239	687	47 19 69	26.211
638	40 70 44	25.259	688	47 33 44	26.230
639	40 83 21	25.278	689	47 47 21	26.249
640	40 96 00	25.298	690	47 61 00	26.268
641	41 08 81	25.318	691	47 74 81	26.287
642	41 21 64	25.338	692	47 88 64	26.306
643	41 34 49	25.357	693	48 02 49	26.325
644	41 47 36	25.377	694	48 16 36	26.344
645	41 60 25	25.397	695	48 30 25	26.363
646	41 73 16	25.417	696	48 44 16	26.382
647	41 86 09	25.436	697	48 58 09	26.401
648	41 99 04	25.456	698	48 72 04	26.420
649	42 12 01	25.475	699	48 86 01	26.439
650	42 25 00	25.495	700	49 00 00	26.458

Number	Square	Square root	Number	Square	Square root
701	49 14 01	26.476	751	56 40 01	27.404
702	49 28 04	26.495	752	56 55 04	27.423
703	49 42 09	26.514	753	56 70 09	27.441
704	49 56 16	26.533	754	56 85 16	27.459
705	49 70 25	26.552	755	57 00 25	27.477
706	49 84 36	26.571	756	57 15 36	27.495
707	49 98 49	26.589	757	57 30 49	27.514
708	50 12 64	26.608	758	57 45 64	27.532
709	50 26 81	26.627	759	57 60 81	27.550
710	50 41 00	26.646	760	57 76 00	27.568
711	50 55 21	26.665	761	57 91 21	27.586
712	50 69 44	26.683	762	58 06 44	27.604
713	50 83 69	26.702	763	58 21 69	27.622
714	50 97 96	26.721	764	58 36 96	27.641
715	51 12 25	26.739	765	58 52 25	27.659
716	51 26 56	26.758	766	58 67 56	27.677
717	51 40 89	26.777	767	58 82 89	27.695
718	51 55 24	26.796	768	58 98 24	27.713
719	51 69 61	26.814	769	59 13 61	27.731
720	51 84 00	26.833	770	59 29 00	27.749
721	51 98 41	26.851	771	59 44 41	27.767
722	52 12 84	26.870	772	59 59 84	27.785
723	52 27 29	26.889	773	59 75 29	27.803
724	52 41 76	26.907	774	59 90 76	27.821
725	52 56 25	26.926	775	60 06 25	27.839
726	52 70 76	26.944	776	60 21 76	27.857
727	52 85 29	26.963	777	60 37 29	27.875
728	52 99 84	26.981	778	60 52 84	27.893
729	53 14 41	27.000	779	60 68 41	27.911
730	53 29 00	27.019	780	60 84 00	27.928
731	53 43 61	27.037	781	60 99 61	27.946
732	53 58 24	27.055	782	61 15 24	27.964
733	53 72 89	27.074	783	61 30 89	27.982
734	53 87 56	27.092	784	61 46 56	28.000
735	54 02 25	27.111	785	61 62 25	28.018
736	54 16 96	27.129	786	61 77 96	28.036
737	54 31 69	27.148	787	61 93 69	28.054
738	54 46 44	27.166	788	62 09 44	28.071
739	54 61 21	27.185	789	62 25 21	28.089
740	54 76 00	27.203	790	62 41 00	28.107
741	54 90 81	27.221	791	62 56 81	28.125
742	55 05 64	27.240	792	62 72 64	28.142
743	55 20 49	27.258	793	62 88 49	28.160
744	55 35 36	27.276	794	63 04 36	28.178
745	55 50 25	27.295	795	63 20 25	28.196
746	55 65 16	27.313	796	63 36 16	28.213
747	55 80 09	27.331	797	63 52 09	28.231
748	55 95 04	27.350	798	63 68 04	28.249
749	56 10 01	27.368	799	63 84 01	28.267
750	56 25 00	27.386	800	64 00 00	28.284

Number	Square	Square root	Number	Square	Square root
801	64 16 01	28.302	851	72 42 01	29.172
802	64 32 04	28.320	852	72 59 04	29.189
803	64 48 09	28.337	853	72 76 09	29.206
804	64 64 16	28.355	854	72 93 16	29.223
805	64 80 25	28.373	855	73 10 25	29.240
806	64 96 36	28.390	856	73 27 36	29.257
807	65 12 49	28.408	857	73 44 49	29.275
808	65 28 64	28.425	858	73 61 64	29.292
809	65 44 81	28.443	859	73 78 81	29.309
810	65 61 00	28.460	860	73 96 00	29.326
811	65 77 21	28.478	861	74 13 21	29.343
812	65 93 44	28.496	862	74 30 44	29.360
813	66 09 69	28.513	863	74 47 69	29.377
814	66 25 96	28.531	854	74 64 96	29.394
815	66 42 25	28.548	865	74 82 25	29.411
816	66 58 56	28.566	866	74 99 56	29.428
817	66 74 89	28.583	867	75 16 89	29.445
818	66 91 24	28.601	868	75 34 24	29.462
819	67 07 61	28.618	869	75 51 61	29.479
820	67 24 00	28.636	870	75 69 00	29.496
821	67 40 41	28.653	871	75 86 41	29.513
822	67 56 84	28.671	872	76 03 84	29.530
823	67 73 29	28.688	873	76 21 29	29.547
824	67 89 76	28.705	874	76 38 76	29.563
825	68 06 25	28.723	875	76 56 25	29.580
826	68 22 76	28.740	876	76 73 76	29.597
827	68 39 29	28.758	877	76 91 29	29.614
828	68 55 84	28.775	878	77 08 84	29.631
829	68 72 41	28.792	879	77 26 41	29.648
830	68 89 00	28.810	880	77 44 00	29.665
831	69 05 61	28.827	881	77 61 61	29.682
832	69 22 24	28.844	882	77 79 24	29.698
833	69 38 89	28.862	883	77 96 89	29.715
834	69 55 56	28.879	884	78 14 56	29.732
835	69 72 25	28.896	885	78 32 25	29.749
836	69 88 96	28.914	886	78 49 96	29.766
837	70 05 69	28.931	887	78 67 69	29.783
838	70 22 44	28.948	888	78 85 44	29.799
839	70 39 21	28.965	889	79 03 21	29.816
840	70 56 00	28.983	890	79 21 00	29.833
841	70 72 81	29.000	891	79 38 81	29.850
842	70 89 64	29.017	892	79 56 64	29.866
843	71 06 49	29.034	893	79 74 49	29.883
844	71 23 36	29.052	894	79 92 36	29.900
845	71 40 25	29.069	895	80 10 25	29.916
846	71 57 16	29.086	896	80 28 16	29.933
847	71 74 09	29.103	897	80 46 09	29.950
848	71 91 04	29.120	898	80 64 04	29.967
849	72 08 01	29.138	899	80 82 01	29.983
850	72 25 00	29.155	900	81 00 00	30.000

Number	Square	Square root	Number	Square	Square root
901	81 18 01	30.017	951	90 44 01	30.838
902	81 36 04	30.033	952	90 63 04	30.854
903	81 54 09	30.050	953	90 82 09	30.871
904	81 72 16	30.067	954	91 01 16	30.887
905	81 90 25	30.083	955	91 20 25	30.903
906	82 08 36	30.100	956	91 39 36	30.919
907	82 26 49	30.116	957	91 58 49	30.935
908	82 44 64	30.133	958	91 77 64	30.952
909	82 62 81	30.150	959	91 96 81	30.968
910	82 81 00	30.166	960	92 16 00	30.984
911	82 99 21	30.183	961	92 35 21	31.000
912	83 17 44	30.199	962	92 54 44	31.016
913	83 35 69	30.216	963	92 73 69	31.032
914	83 53 96	30.232	964	92 92 96	31.048
915	83 72 25	30.249	965	93 12 25	31.064
916	83 90 56	30.265	966	93 31 56	31.081
917	84 08 89	30.282	967	93 50 89	31.097
918	84 27 24	30.299	968	93 70 24	31.113
919	84 45 61	30.315	969	93 89 61	31.129
920	84 64 00	30.332	970	94 09 00	31.145
921	84 82 41	30.348	971	94 28 41	31.161
922	85 00 84	30.364	972	94 47 84	31.177
923	85 19 29	30.381	973	94 67 29	31.193
924	85 37 76	30.397	974	94 86 76	31.209
925	85 56 25	30.414	975	95 06 25	31.225
926	85 74 76	30.430	976	95 25 76	31.241
927	85 93 29	30.447	977	95 45 29	31.257
928	86 11 84	30.463	978	95 64 84	31.273
929	86 30 41	30.480	979	95 84 41	31.289
930	86 49 00	30.496	980	96 04 00	31.305
931	86 67 61	30.512	981	96 23 61	31.321
932	86 86 24	30.529	982	96 43 24	31.337
933	87 04 89	30.545	983	96 62 89	31.353
934	87 23 56	30.561	984	96 82 56	31.369
935	87 42 25	30.578	985	97 02 25	31.385
936	87 60 96	30.594	986	97 21 96	31.401
937	87 79 69	30.610	987	97 41 69	31.417
938	87 98 44	30.627	988	97 61 44	31.432
939	88 17 21	30.643	989	97 81 21	31.448
940	88 36 00	30.659	990	98 01 00	31.464
941	88 54 81	30.676	991	98 20 81	31.480
942	88 73 64	30.692	992	98 40 64	31.496
943	88 92 49	30.708	993	98 60 49	31.512
944	89 11 36	30.725	994	98 80 36	31.528
945	89 30 25	30.741	995	99 00 25	31.544
946	89 49 16	30.757	996	99 20 16	31.559
947	89 68 09	30.773	997	99 40 09	31.575
948	89 87 04	30.790	998	99 60 04	31.591
949	90 06 01	30.806	999	99 80 01	31.607
950	90 25 00	30.822	1000	100 00 00	31.623

INTRODUCTION TO STATISTICS

A.16. Random numbers.

```
10 09 73 25 33    76 52 01 35 86    34 67 35 48 76    80 95 90 91 17    39 29 27 49 45
37 54 20 48 05    64 89 47 42 96    24 80 52 40 37    20 63 61 04 02    00 82 29 16 65
08 42 26 89 53    19 64 50 93 03    23 20 90 25 60    15 95 33 47 64    35 08 03 36 06
99 01 90 25 29    09 37 67 07 15    38 31 13 11 65    88 67 67 43 97    04 43 62 76 59
12 80 79 99 70    80 15 73 61 47    64 03 23 66 53    98 95 11 68 77    12 17 17 68 33

66 06 57 47 17    34 07 27 68 50    36 69 73 61 70    65 81 33 98 85    11 19 92 91 70
31 06 01 08 05    45 57 18 24 06    35 30 34 26 14    86 79 90 74 39    23 40 30 97 32
85 26 97 76 02    02 05 16 56 92    68 66 57 48 18    73 05 38 52 47    18 62 38 85 79
63 57 33 21 35    05 32 54 70 48    90 55 35 75 48    28 46 82 87 09    83 49 12 56 24
73 79 64 57 53    03 52 96 47 78    35 80 83 42 82    60 93 52 03 44    35 27 38 84 35

98 52 01 77 67    14 90 56 86 07    22 10 94 05 58    60 97 09 34 33    50 50 07 39 98
11 80 50 54 31    39 80 82 77 32    50 72 56 82 48    29 40 52 42 01    52 77 56 78 51
83 45 29 96 34    06 28 89 80 83    13 74 67 00 78    18 47 54 06 10    68 71 17 78 17
88 68 54 02 00    86 50 75 84 01    36 76 66 79 51    90 36 47 64 93    29 60 91 10 62
99 59 46 73 48    87 51 76 49 69    91 82 60 89 28    93 78 56 13 68    23 47 83 41 13

65 48 11 76 74    17 46 85 09 50    58 04 77 69 74    73 03 95 71 86    40 21 81 65 44
80 12 43 56 35    17 72 70 80 15    45 31 82 23 74    21 11 57 82 53    14 38 55 37 63
74 35 09 98 17    77 40 27 72 14    43 23 60 02 10    45 52 16 42 37    96 28 60 26 55
69 91 62 68 03    66 25 22 91 48    36 93 68 72 03    76 62 11 39 90    94 40 05 64 18
09 89 32 05 05    14 22 56 85 14    46 42 75 67 88    96 29 77 88 22    54 38 21 45 98

91 49 91 45 23    68 47 92 76 86    46 16 28 35 54    94 75 08 99 23    37 08 92 00 48
80 33 69 45 98    26 94 03 68 58    70 29 73 41 35    53 14 03 33 40    42 05 08 23 41
44 10 48 19 49    85 15 74 79 54    32 97 92 65 75    57 60 04 08 81    22 22 20 64 13
12 55 07 37 42    11 10 00 20 40    12 86 07 46 97    96 64 48 94 39    28 70 72 58 15
63 60 64 93 29    16 50 53 44 84    40 21 95 25 63    43 65 17 70 82    07 20 73 17 90

61 19 69 04 46    26 45 74 77 74    51 92 43 37 29    65 39 45 95 93    42 58 26 05 27
15 47 44 52 66    95 27 07 99 53    59 36 78 38 48    82 39 61 01 18    33 21 15 94 66
94 55 72 85 73    67 89 75 43 87    54 62 24 44 31    91 19 04 25 92    92 92 74 59 73
42 48 11 62 13    97 34 40 87 21    16 86 84 87 67    03 07 11 20 59    25 70 14 66 70
23 52 37 83 17    73 20 88 98 37    68 93 59 14 16    26 25 22 96 63    05 52 28 25 62

04 49 35 24 94    75 24 63 38 24    45 86 25 10 25    61 96 27 93 35    65 33 71 24 72
00 54 99 76 54    64 05 18 81 59    96 11 96 38 96    54 69 28 23 91    23 28 72 95 29
35 96 31 53 07    26 89 80 93 54    33 35 13 54 62    77 97 45 00 24    90 10 33 93 33
59 80 80 83 91    45 42 72 68 42    83 60 94 97 00    13 02 12 48 92    78 56 52 01 06
46 05 88 52 36    01 39 09 22 86    77 28 14 40 77    93 91 08 36 47    70 61 74 29 41

32 17 90 05 97    87 37 92 52 41    05 56 70 70 07    86 74 31 71 57    85 39 41 18 38
69 23 46 14 06    20 11 74 52 04    15 95 66 00 00    18 74 39 24 23    97 11 89 63 38
19 56 54 14 30    01 75 87 53 79    40 41 92 15 85    66 67 43 68 06    84 96 28 52 07
45 15 51 49 38    19 47 60 72 46    43 66 79 45 43    59 04 79 00 33    20 82 66 95 41
94 86 43 19 94    36 16 81 08 51    34 88 88 15 53    01 54 03 54 56    05 01 45 11 76

98 08 62 48 26    45 24 02 84 04    44 99 90 88 96    39 09 47 34 07    35 44 13 18 80
33 18 51 62 32    41 94 15 09 49    89 43 54 85 81    88 69 54 19 94    37 54 87 30 43
80 95 10 04 06    96 38 27 07 74    20 15 12 33 87    25 01 62 52 98    94 62 46 11 71
79 75 24 91 40    71 96 12 82 96    69 86 10 25 91    74 85 22 05 39    00 38 75 95 79
18 63 33 25 37    98 14 50 65 71    31 01 02 46 74    05 45 56 14 27    77 93 89 19 36

74 02 94 39 02    77 55 73 22 70    97 79 01 71 19    52 52 75 80 21    80 81 45 17 48
54 17 84 56 11    80 99 33 71 43    05 33 51 29 69    56 12 71 92 55    36 04 09 03 24
11 66 44 98 83    52 07 98 48 27    59 38 17 15 39    09 97 33 34 40    88 46 12 33 56
48 32 47 79 28    31 24 96 47 10    02 29 53 68 70    32 30 75 75 46    15 02 00 99 94
69 07 49 41 38    87 63 79 19 76    35 58 40 44 01    10 51 82 16 15    01 84 87 69 38
```

This table is reproduced by permission of the RAND Corporation.

```
09 18 82 00 97   32 82 53 95 27   04 22 08 63 04   83 38 98 73 74   64 27 85 80 44
90 04 58 54 97   51 98 15 06 54   94 93 88 19 97   91 87 07 61 50   68 47 66 46 59
73 18 95 02 07   47 67 72 62 69   62 29 06 44 64   27 12 46 70 18   41 36 18 27 60
75 76 87 64 90   20 97 18 17 49   90 42 91 22 72   95 37 50 58 71   93 82 34 31 78
54 01 64 40 56   66 28 13 10 03   00 68 22 73 98   20 71 45 32 95   07 70 61 78 13

08 35 86 99 10   78 54 24 27 85   13 66 15 88 73   04 61 89 75 53   31 22 30 84 20
28 30 60 32 64   81 33 31 05 91   40 51 00 78 93   32 60 46 04 75   94 11 90 18 40
53 84 08 62 33   81 59 41 36 28   51 21 59 02 90   28 46 66 87 95   77 76 22 07 91
91 75 75 37 41   61 61 36 22 69   50 26 39 02 12   55 78 17 65 14   83 48 34 70 55
89 41 59 26 94   00 39 75 83 91   12 60 71 76 46   48 94 97 23 06   94 54 13 74 08

77 51 30 38 20   86 83 42 99 01   68 41 48 27 74   51 90 81 39 80   72 89 35 55 07
19 50 23 71 74   69 97 92 02 88   55 21 02 97 73   74 28 77 52 51   65 34 46 74 15
21 81 85 93 13   93 27 88 17 57   05 68 67 31 56   07 08 28 50 46   31 85 33 84 52
51 47 46 64 99   68 10 72 36 21   94 04 99 13 45   42 83 60 91 91   08 00 74 54 49
99 55 96 83 31   62 53 52 41 70   69 77 71 28 30   74 81 97 81 42   43 86 07 28 34

33 71 34 80 07   93 58 47 28 69   51 92 66 47 21   58 30 32 98 22   93 17 49 39 72
85 27 48 68 93   11 30 32 92 70   28 83 43 41 37   73 51 59 04 00   71 14 84 36 43
84 13 38 96 40   44 03 55 21 66   73 85 27 00 91   61 22 26 05 61   62 32 71 84 23
56 73 21 62 34   17 39 59 61 31   10 12 39 16 22   85 49 65 75 60   81 60 41 88 80
65 13 85 68 06   87 64 88 52 61   34 31 36 58 61   45 87 52 10 69   85 64 44 72 77

38 00 10 21 76   81 71 91 17 11   71 60 29 29 37   74 21 96 40 49   65 58 44 96 98
37 40 29 63 97   01 30 47 75 86   56 27 11 00 86   47 32 46 26 05   40 03 03 74 38
97 12 54 03 48   87 08 33 14 17   21 81 53 92 50   75 23 76 20 47   15 50 12 95 78
21 82 64 11 34   47 14 33 40 72   64 63 88 59 02   49 13 90 64 41   03 85 65 45 52
73 13 54 27 42   95 71 90 90 35   85 79 47 42 96   08 78 98 81 56   64 69 11 92 02

07 63 87 79 29   03 06 11 80 72   96 20 74 41 56   23 82 19 95 38   04 71 36 69 94
60 52 88 34 41   07 95 41 98 14   59 17 52 06 95   05 53 35 21 39   61 21 20 64 55
83 59 63 56 55   06 95 89 29 83   05 12 80 97 19   77 43 35 37 83   92 30 15 04 98
10 85 06 27 46   99 59 91 05 07   13 49 90 63 19   53 07 57 18 39   06 41 01 93 62
39 82 09 89 52   43 62 26 31 47   64 42 18 08 14   43 80 00 93 51   31 02 47 31 67

59 58 00 64 78   75 56 97 88 00   88 83 55 44 86   23 76 80 61 56   04 11 10 84 08
38 50 80 73 41   23 79 34 87 63   90 82 29 70 22   17 71 90 42 07   95 95 44 99 53
30 69 27 06 68   94 68 81 61 27   56 19 68 00 91   82 06 76 34 00   05 46 26 92 00
65 44 39 56 59   18 28 82 74 37   49 63 22 40 41   08 33 76 56 76   96 29 99 08 36
27 26 75 02 64   13 19 27 22 94   07 47 74 46 06   17 98 54 89 11   97 34 13 03 58

91 30 70 69 91   19 07 22 42 10   36 69 95 37 28   28 82 53 57 93   28 97 66 62 52
68 43 49 46 88   84 47 31 36 22   62 12 69 84 08   12 84 38 25 90   09 81 59 31 46
48 90 81 58 77   54 74 52 45 91   35 70 00 47 54   83 82 45 26 92   54 13 05 51 60
06 91 34 51 97   42 67 27 86 01   11 88 30 95 28   63 01 19 89 01   14 97 44 03 44
10 45 51 60 19   14 21 03 37 12   91 34 23 78 21   88 32 58 08 51   43 66 77 08 83

12 88 39 73 43   65 02 76 11 84   04 28 50 13 92   17 97 41 50 77   90 71 22 67 69
21 77 83 09 76   38 80 73 69 61   31 64 94 20 96   63 28 10 20 23   08 81 64 74 49
19 52 35 95 15   65 12 25 96 59   86 28 36 82 58   69 57 21 37 98   16 43 59 15 29
67 24 55 26 70   35 58 31 65 63   79 24 68 66 86   76 46 33 42 22   26 65 59 08 02
60 58 44 73 77   07 50 03 79 92   45 13 42 65 29   26 76 08 36 37   41 32 64 43 44

53 85 34 13 77   36 06 69 48 50   58 83 87 38 59   49 36 47 33 31   96 24 04 36 42
24 63 73 87 36   74 38 48 93 42   52 62 30 79 92   12 36 91 86 01   03 74 28 38 73
83 08 01 24 51   38 99 22 28 15   07 75 95 17 77   97 37 72 75 85   51 97 23 78 67
16 44 42 43 34   36 15 19 90 73   27 49 37 09 39   85 13 03 25 52   54 84 65 47 59
60 79 01 81 57   57 17 86 57 62   11 16 17 85 76   45 81 95 29 79   65 13 00 48 60
```

```
03 99 11 04 61    93 71 61 68 94    66 08 32 46 53    84 60 95 82 32    88 61 81 91 61
38 55 59 55 54    32 88 65 97 80    08 35 56 08 60    29 73 54 77 62    71 29 92 38 53
17 54 67 37 04    92 05 24 62 15    55 12 12 92 81    59 07 60 79 36    27 95 45 89 09
32 64 35 28 61    95 81 90 68 31    00 91 19 89 36    76 35 59 37 79    80 86 30 05 14
69 57 26 87 77    39 51 03 59 05    14 06 04 06 19    29 54 96 96 16    33 56 46 07 80

24 12 26 65 91    27 69 90 64 94    14 84 54 66 72    61 95 87 71 00    90 89 97 57 54
61 19 63 02 31    92 96 26 17 73    41 83 95 53 82    17 26 77 09 43    78 03 87 02 67
30 53 22 17 04    10 27 41 22 02    39 68 52 33 09    10 06 16 88 29    55 98 66 64 85
03 78 89 75 99    75 86 72 07 17    74 41 65 31 66    35 20 83 33 74    87 53 90 88 23
48 22 86 33 79    85 78 34 76 19    53 15 26 74 33    35 66 35 29 72    16 81 86 03 11

60 36 59 46 53    35 07 53 39 49    42 61 42 92 97    01 91 82 83 16    98 95 37 32 31
83 79 94 24 02    56 62 33 44 42    34 99 44 13 74    70 07 11 47 36    09 95 81 80 65
32 96 00 74 05    36 40 98 32 32    99 38 54 16 00    11 13 30 75 86    15 91 70 62 53
19 32 25 38 45    57 62 05 26 06    66 49 76 86 46    78 13 86 65 59    19 64 09 94 13
11 22 09 47 47    07 39 93 74 08    48 50 92 39 29    27 48 24 54 76    85 24 43 51 59

31 75 15 72 60    68 98 00 53 39    15 47 04 83 55    88 65 12 25 96    03 15 21 91 21
88 49 29 93 82    14 45 40 45 04    20 09 49 89 77    74 84 39 34 13    22 10 97 85 08
30 93 44 77 44    07 48 18 38 28    73 78 80 65 33    28 59 72 04 05    94 20 52 03 80
22 88 84 88 93    27 49 99 87 48    60 53 04 51 28    74 02 28 46 17    82 03 71 02 68
78 21 21 69 93    35 90 29 13 86    44 37 21 54 86    65 74 11 40 14    87 48 13 72 20

41 84 98 45 47    46 85 05 23 26    34 67 75 83 00    74 91 06 43 45    19 32 58 15 49
46 35 23 30 49    69 24 89 34 60    45 30 50 75 21    61 31 83 18 55    14 41 37 09 51
11 08 79 62 94    14 01 33 17 92    59 74 76 72 77    76 50 33 45 13    39 66 37 75 44
52 70 10 83 37    56 30 38 73 15    16 52 06 96 76    11 65 49 98 93    02 18 16 81 61
57 27 53 68 98    81 30 44 85 85    68 65 22 73 76    92 85 25 58 66    88 44 80 35 84

20 85 77 31 56    70 28 42 43 26    79 37 59 52 20    01 15 96 32 67    10 62 24 83 91
15 63 38 49 24    90 41 59 36 14    33 52 12 66 65    55 82 34 76 41    86 22 53 17 04
92 69 44 82 97    39 90 40 21 15    59 58 94 90 67    66 82 14 15 75    49 76 70 40 37
77 61 31 90 19    88 15 20 00 80    20 55 49 14 09    96 27 74 82 57    50 81 69 76 16
38 68 83 24 86    45 13 46 35 45    59 40 47 20 59    43 94 75 16 80    43 85 25 96 93

25 16 30 18 89    70 01 41 50 21    41 29 06 73 12    71 85 71 59 57    68 97 11 14 03
65 25 10 76 29    37 23 93 32 95    05 87 00 11 19    92 78 42 63 40    18 47 76 56 22
36 81 54 36 25    18 63 73 75 09    82 44 49 90 05    04 92 17 37 01    14 70 79 39 97
64 39 71 16 92    05 32 78 21 62    20 24 78 17 59    45 19 72 53 32    83 74 52 25 67
04 51 52 56 24    95 09 66 79 46    48 46 08 55 58    15 19 11 87 82    16 93 03 33 61

83 76 16 08 73    43 25 38 41 45    60 83 32 59 83    01 29 14 13 49    20 36 80 71 26
14 38 70 63 45    80 85 40 92 79    43 52 90 63 18    38 38 47 47 61    41 19 63 74 80
51 32 19 22 46    80 08 87 70 74    88 72 25 67 36    66 16 44 94 31    66 91 93 16 78
72 47 20 00 08    80 89 01 80 02    94 81 33 19 00    54 15 58 34 36    35 35 25 41 31
05 46 65 53 06    93 12 81 84 64    74 45 79 05 61    72 84 81 18 34    79 98 26 84 16

39 52 87 24 84    82 47 42 55 93    48 54 53 52 47    18 61 91 36 74    18 61 11 92 41
81 61 61 87 11    53 34 24 42 76    75 12 21 17 24    74 62 77 37 07    58 31 91 59 97
07 58 61 61 20    82 64 12 28 20    92 90 41 31 41    32 39 21 97 63    61 19 96 79 40
90 76 70 42 35    13 57 41 72 00    69 90 26 37 42    78 46 42 25 01    18 62 79 08 72
40 18 82 81 93    29 59 38 86 27    94 97 21 15 98    62 09 53 67 87    00 44 15 89 97

34 41 48 21 57    86 88 75 50 87    19 15 20 00 23    12 30 28 07 83    32 62 46 86 91
63 43 97 53 63    44 98 91 68 22    36 02 40 08 67    76 37 84 16 05    65 96 17 34 88
67 04 90 90 70    93 39 94 55 47    94 45 87 42 84    05 04 14 98 07    20 28 83 40 60
79 49 50 41 46    52 16 29 02 86    54 15 83 42 43    46 97 83 54 82    59 36 29 59 38
91 70 43 05 52    04 73 72 10 31    75 05 19 30 29    47 66 56 43 82    99 78 29 34 78
```

```
94 01 54 68 74    32 44 44 82 77    59 82 09 61 63    64 65 42 58 43    41 14 54 28 20
74 10 88 82 22    88 57 07 40 15    25 70 49 10 35    01 75 51 47 50    48 96 83 86 03
62 88 08 78 73    95 16 05 92 21    22 30 49 03 14    72 87 71 73 34    39 28 30 41 49
11 74 81 21 02    80 58 04 18 67    17 71 05 96 21    06 55 40 78 50    73 95 07 95 52
17 94 40 56 00    60 47 80 33 43    25 85 25 89 05    57 21 63 96 18    49 85 69 93 26

66 06 74 27 92    95 04 35 26 80    46 78 05 64 87    09 97 15 94 81    37 00 62 21 86
54 24 49 10 30    45 54 77 08 18    59 84 99 61 69    61 45 92 16 47    87 41 71 71 98
30 94 55 75 89    31 73 25 72 60    47 67 00 76 54    46 37 62 53 66    94 74 64 95 80
69 17 03 74 03    86 99 59 03 07    94 30 47 18 03    26 82 50 55 11    12 45 99 13 14
08 34 58 89 75    35 84 18 57 71    08 10 55 99 87    87 11 22 14 76    14 71 37 11 81

27 76 74 35 84    85 30 18 89 77    29 49 06 97 14    73 03 54 12 07    74 69 90 93 10
13 02 51 43 38    54 06 61 52 43    47 72 46 67 33    47 43 14 39 05    31 04 85 66 99
80 21 73 62 92    98 52 52 43 35    24 43 22 48 96    43 27 75 88 74    11 46 61 60 82
10 87 56 20 04    90 39 16 11 05    57 41 10 63 68    53 85 63 07 43    08 67 08 47 41
54 12 75 73 26    26 62 91 90 87    24 47 28 87 79    30 54 02 78 86    61 73 27 54 54

60 31 14 28 24    37 30 14 26 78    45 99 04 32 42    17 37 45 20 03    70 70 77 02 14
49 73 97 14 84    92 00 39 80 86    76 66 87 32 09    59 20 21 19 73    02 90 23 32 50
78 62 65 15 94    16 45 39 46 14    39 01 49 70 66    83 01 20 98 32    25 57 17 76 28
66 69 21 39 86    99 83 70 05 82    81 23 24 49 87    09 50 49 64 12    90 19 37 95 68
44 07 12 80 91    07 36 29 77 03    76 44 74 25 37    98 52 49 78 31    65 70 40 95 14

41 46 88 51 49    49 55 41 79 94    14 92 43 96 50    95 29 40 05 56    70 48 10 69 05
94 55 93 75 59    49 67 85 31 19    70 31 20 56 82    66 98 63 40 99    74 47 42 07 40
41 61 57 03 60    64 11 45 86 60    90 85 06 46 18    80 62 05 17 90    11 43 63 80 72
50 27 39 31 13    41 79 48 68 61    24 78 18 96 83    55 41 18 56 67    77 53 59 98 92
41 39 68 05 04    90 67 00 82 89    40 90 20 50 69    95 08 30 67 83    28 10 25 78 16

25 80 72 42 60    71 52 97 89 20    72 68 20 73 85    90 72 65 71 66    98 88 40 85 83
06 17 09 79 65    88 30 29 80 41    21 44 34 18 08    68 98 48 36 20    89 74 79 88 82
60 80 85 44 44    74 41 28 11 05    01 17 62 88 38    36 42 11 64 89    18 05 95 10 61
80 94 04 48 93    10 40 83 62 22    80 58 27 19 44    92 63 84 03 33    67 05 41 60 67
19 51 69 01 20    46 75 97 16 43    13 17 75 52 92    21 03 68 28 08    77 50 19 74 27

49 38 65 44 80    23 60 42 35 54    21 78 54 11 01    91 17 81 01 74    29 42 09 04 38
06 31 28 89 40    15 99 56 93 21    47 45 86 48 09    98 18 98 18 51    29 65 18 42 15
60 94 20 03 07    11 89 79 26 74    40 40 56 80 32    96 71 75 42 44    10 70 14 13 93
92 32 99 89 32    78 28 44 63 47    71 20 99 20 61    39 44 89 31 36    25 72 20 85 64
77 93 66 35 74    31 38 45 19 24    85 56 12 96 71    58 13 71 78 20    22 75 13 65 18

38 10 17 77 56    11 65 71 38 97    95 88 95 70 67    47 64 81 38 85    70 66 99 34 06
39 64 16 94 57    91 33 92 25 02    92 61 38 97 19    11 94 75 62 03    19 32 42 05 04
84 05 44 04 55    99 39 66 36 80    67 66 76 06 31    69 18 19 68 45    38 52 51 16 00
47 46 80 35 77    57 64 96 32 66    24 70 07 15 94    14 00 42 31 53    69 24 90 57 47
43 32 13 13 70    28 97 72 38 96    76 47 96 85 62    62 34 20 75 89    08 89 90 59 85

64 28 16 18 26    18 55 56 49 37    13 17 33 33 65    78 85 11 64 99    87 06 41 30 75
66 84 77 04 95    32 35 00 29 85    86 71 63 87 46    26 31 37 74 63    55 38 77 26 81
72 46 13 32 30    21 52 95 34 24    92 58 10 22 62    78 43 86 62 76    18 39 67 35 38
21 03 29 10 50    13 05 81 62 18    12 47 05 65 00    15 29 27 61 39    59 52 65 21 13
95 36 26 70 11    06 65 11 61 36    01 01 60 08 57    55 01 85 63 74    35 82 47 17 08

49 71 29 73 80    10 40 45 54 52    34 03 06 07 26    75 21 11 02 71    36 63 36 84 24
58 27 56 17 64    97 58 65 47 16    50 25 94 63 45    87 19 54 60 92    26 78 76 09 39
89 51 41 17 88    68 22 42 34 17    73 95 97 61 45    30 34 24 02 77    11 04 97 20 49
15 47 25 06 69    48 13 93 67 32    46 87 43 70 88    73 46 50 98 19    58 86 93 52 20
12 12 08 61 24    51 24 74 43 02    60 88 35 21 09    21 43 73 67 86    49 22 67 78 37
```

Table A.17. Four place common logarithms of numbers.

	0	1	2	3	4	5	6	7	8	9
10	·0000	0043	0086	0128	0170	0212	0253	0294	0334	0374
11	·0414	0453	0492	0531	0569	0607	0645	0682	0719	0755
12	·0792	0828	0864	0899	0934	0969	1004	1038	1072	1106
13	·1139	1173	1206	1239	1271	1303	1335	1367	1399	1430
14	·1461	1492	1523	1553	1584	1614	1644	1673	1703	1732
15	·1761	1790	1818	1847	1875	1903	1931	1959	1987	2014
16	·2041	2068	2095	2122	2148	2175	2201	2227	2253	2279
17	·2304	2330	2355	2380	2405	2430	2455	2480	2504	2529
18	·2553	2577	2601	2625	2648	2672	2695	2718	2742	2765
19	·2788	2810	2833	2856	2878	2900	2923	2945	2967	2989
20	·3010	3032	3054	3075	3096	3118	3139	3160	3181	3201
21	·3222	3243	3263	3284	3304	3324	3345	3365	3385	3404
22	·3424	3444	3464	3483	3502	3522	3541	3560	3579	3598
23	·3617	3636	3655	3674	3692	3711	3729	3747	3766	3784
24	·3802	3820	3838	3856	3874	3892	3909	3927	3945	3962
25	·3979	3997	4014	4031	4048	4065	4082	4099	4116	4133
26	·4150	4166	4183	4200	4216	4232	4249	4265	4281	4298
27	·4314	4330	4346	4362	4378	4393	4409	4425	4440	4456
28	·4472	4487	4502	4518	4533	4548	4564	4579	4594	4609
29	·4624	4639	4654	4669	4603	4698	4713	4728	4742	4757
30	·4771	4786	4800	4814	4829	4843	4857	4871	4886	4900
31	·4914	4928	4942	4955	4969	4983	4997	5011	5024	5038
32	·5051	5065	5079	5092	5105	5119	5132	5145	5159	5172
33	·5185	5198	5211	5224	5237	5250	5263	5276	5289	5302
34	·5315	5328	5340	5353	5366	5378	5391	5403	5416	5428
35	·5441	5453	5465	5478	5490	5502	5514	5527	5539	5551
36	·5563	5575	5587	5599	5611	5623	5635	5647	5658	5670
37	·5682	5694	5705	5717	5729	5740	5752	5763	5775	5786
38	·5798	5809	5821	5832	5843	5855	5866	5877	5888	5899
39	·5911	5922	5933	5944	5955	5966	5977	5988	5999	6010
40	·6021	6031	6042	6053	6064	6075	6085	6096	6107	6117
41	·6128	6138	6149	6160	61·70	6180	6191	6201	6212	6222
42	·6232	6243	6253	6263	6274	6284	6294	6304	6314	6325
43	·6335	6345	6355	6365	6375	6385	6395	6405	6415	6425
44	·6435	6444	6454	6464	6474	6484	6493	6503	6513	6522
45	·6532	6542	6551	6561	6571	6580	6590	6599	6609	6618
46	·6628	6637	6646	6656	6665	6675	6684	6693	6702	6712
47	·6721	6730	6739	6749	6758	6767	6776	6785	6794	6803
48	·6812	6821	6830	6839	6848	6857	6866	6875	6884	6893
49	·6902	6911	6920	6928	6937	6946	6955	6964	6972	6981
50	·6990	6998	7007	7016	7024	7033	7042	7050	7059	7067
51	·7076	7084	7093	7101	7110	7118	7126	7135	7143	7152
52	·7160	7168	7177	7185	7193	7202	7210	7218	7226	7235
53	·7243	7251	7259	7267	7275	7284	7292	7300	7308	7316
54	·7324	7332	7340	7348	7356	7364	7372	7380	7388	7396

	0	1	2	3	4	5	6	7	8	9
55	·7404	7412	7419	7427	7435	7443	7451	7459	7466	7474
56	·7482	7490	7497	7505	7513	7520	7528	7536	7543	7551
57	·7559	7566	7574	7582	7589	7597	7604	7612	7619	7627
58	·7634	7642	7649	7657	7664	7672	7679	7686	7694	7701
59	·7709	7716	7723	7731	7738	7745	7752	7760	7767	7774
60	·7782	7789	7796	7803	7810	7818	7825	7832	7839	7846
61	·7853	7860	7868	7875	7882	7889	7896	7903	7910	7917
62	·7924	7931	7938	7945	7952	7959	7966	7973	7980	7987
63	·7993	8000	8007	8014	8021	8028	8035	8041	8048	8055
64	·8062	8069	8075	8082	8089	8096	8102	8109	8116	8122
65	·8129	8136	8142	8149	8156	8162	8169	8176	8182	8189
66	·8195	8202	8209	8215	8222	8228	8235	8241	8248	8254
67	·8261	8267	8274	8280	8287	8293	8299	8306	8312	8319
68	·8325	8331	8338	8344	8351	8357	8363	8370	8376	8382
69	·8388	8395	8401	8407	8414	8420	8426	8432	8439	8445
70	·8451	8457	8463	8470	8476	8482	8488	8494	8500	8506
71	·8513	8519	8525	8531	8537	8543	8549	8555	8561	8567
72	·8573	8579	8585	8591	8597	8603	8609	8615	8621	8627
73	·8633	8639	8645	8651	8657	8663	8669	8675	8681	8686
74	·8692	8698	8704	8710	8716	8722	8727	8733	8739	8745
75	·8751	8756	8762	8768	8774	8779	8785	8791	8797	8802
76	·8808	8814	8820	8825	8831	8837	8842	8848	8854	8859
77	·8865	8871	8876	8882	8887	8893	8899	8904	8910	8915
78	·8921	8927	8932	8938	8943	8949	8954	8960	8965	8971
79	·8976	8982	8987	8993	8998	9004	9009	9015	9020	9025
80	·9031	9036	9042	9047	9053	9058	9063	9069	9074	9079
81	·9085	9090	9096	9101	9106	9112	9117	9122	9128	9133
82	·9138	9143	9149	9154	9159	9165	9170	9175	9180	9186
83	·9191	9196	9201	9206	9212	9217	9222	9227	9232	9238
84	·9243	9248	9253	9258	9263	9269	9274	9279	9284	9289
85	·9294	9299	9304	9309	9315	9320	9325	9330	9335	9340
86	·9345	9350	9355	9360	9365	9370	9375	9380	9385	9390
87	·9395	9400	9405	9410	9415	9420	9425	9430	9435	9440
88	·9445	9450	9455	9460	9465	9469	9474	9479	9484	9489
89	·9494	9499	9504	9509	9513	9518	9523	9528	9533	9538
90	·9542	9547	9552	9557	9562	9566	9571	9576	9581	9586
91	·9590	9595	9600	9605	9609	9614	9619	9624	9628	9633
92	·9638	9643	9647	9652	9657	9661	9666	9671	9675	9680
93	·9685	9689	9694	9699	9703	9708	9713	9717	9722	9727
94	·9731	9736	9741	9745	9750	9754	9759	9763	9768	9773
95	·9777	9782	9786	9791	9795	9800	9805	9809	9814	9818
96	·9823	9827	9832	9836	9841	9845	9850	9854	9859	9863
97	·9868	9872	9877	9881	9886	9890	9894	9899	9903	9908
98	·9912	9917	9921	9926	9930	9934	9939	9943	9948	9952
99	·9956	9961	9965	9969	9974	9978	9983	9987	9991	9996

Bibliography

Alder, H. L., and E. B. Roessler. *Introduction to Probability and Statistics.* San Francisco: W. H. Freeman and Co., 1964.

Anderson, R. L., and T. A. Bancroft. *Statistical Theory in Research.* New York: McGraw-Hill Book Co., Inc., 1952.

Arkin, H., and R. R. Colton. *Tables for Statisticians.* New York: Barnes & Noble, Inc., 1963.

Armore, S. J. *Introduction to Statistical Analysis and Inference.* New York: John Wiley & Sons, Inc., 1966.

Bernstein, A. L. *A Handbook of Statistics Solutions for the Behavioral Sciences.* New York: Holt, Rinehart and Winston, Inc., 1964.

Black, A. H., N. J. Carlson, and R. L. Solomon. Exploratory Studies of the Conditioning of Autonomic Responses in Curarized Dogs. *Psychological Monographs,* 76 (1962), 1–31.

Blommers, P., and E. F. Lindquist. *Elementary Statistical Methods.* Boston: Houghton Mifflin Co., 1960.

Box, G. E. P. "Non Normality and Tests on Variances." *Biometrika,* (1953), 40 318–35.

Bradley, J. V. "Studies in Research Methodology: I. Compatability of Psychological Measurement with Parametric Assumptions." *Wright Air Development Center Technical Report 58–574–(I)*, Wright-Patterson Air Force Base, Ohio, September 1959.

Bradley, J. V. "Studies in Research Methodology: II. Consequences of Violating Parametric Assumptions." *Wright Air Development Center Technical Report 58–574–(II)*, Wright-Patterson Air Force Base, Ohio, September 1959.

Bradley, J. V. "Studies in Research Methodology: III. Persistence of Sequential Effects Despite Extended Practice." *Aerospace Medical Research Laboratories Technical Report MRL–TDR–62–60*, Wright-Patterson Air Force Base, Ohio, June 1962.

Burch, G. E., and T. Winsor. *A Primer of Electrocardiography,* 3rd Ed. Philadelphia: Lea and Febiger, 1955.

Burch, N. R., and H. E. Childers. "Information Processing in the Time Domain," in W. S. Fields and W. Abbott (eds.), *Information Storage and Neural Control*. Springfield; Ill.: Charles C. Thomas, Publisher, 1963.

Burington, R. S., and D. C. May. *Handbook of Probability and Statistics with Tables*. Sandusky, Ohio: Handbook Publishers, Inc., 1953.

Buros, F. C., and O. K. Buros. *Expressing Educational Measures as Percentile Ranks*. Highland Park, Ill.: The Gryphon Press, 1949.

Cochran, W. G., and G. M. Cox. *Experimental Designs*, 2nd Ed. New York: John Wiley & Sons, Inc., 1957.

Courts, F. A. *Psychological Statistics*. Homewood, Ill.: The Dorsey Press, 1966.

Dixon, W. J., and F. J. Massey. *Introduction to Statistical Analysis*, 2nd Ed. New York: McGraw-Hill Book Co., Inc., 1957.

Downie, N. M. *Fundamentals of Measurement: Techniques and Practices*. New York: Oxford University Press, Inc., 1958.

Downie, N. M., and R. W. Heath. *Basic Statistical Methods*. New York: Harper and Row, Publishers, 1965.

DuBois, P. H. *An Introduction to Psychological Statistics*. New York: Harper and Row, Publishers, 1965.

Dykman, R. A., W. G. Reese, C. R. Galbrecht, and P. G. Thomasson. "Psychophysiological Reactions to Novel Stimuli: Measurement, Adaptation, and Relationship of Psychological and Physiological Variables in the Normal Human." *Annals of the New York Academy of Sciences*, **79** (August 1959), 43–107.

Edwards, A. E. *Expected Values of Discrete Random Variables and Elementary Statistics*. New York: John Wiley & Sons, Inc., 1964.

Edwards, A. L. *Statistical Methods for the Behavioral Sciences*. New York: Rinehart and Co., 1955.

Edwards, A. L. *Experimental Design in Psychological Research*, Rev. Ed. New York: Holt, Rinehart and Winston, 1962.

Edwards, A. L. *Statistical Analysis*. New York: Holt, Rinehart and Winston, 1964.

Granville, W. A. *Elements of the Differential and Integral Calculus*. Boston: Ginn & Co., 1929.

Grings, W. W., and D. E. O'Donnell. "Magnitude of Response to Compounds of Discriminated Stimuli." *Journal of Experimental Psychology*, **52** (1956), 354–59.

Guenther, W. C. *Concepts of Statistical Inference*. New York: McGraw-Hill Book Co. Inc., 1965.

Guilford, J. P. *Fundamental Statistics in Psychology and Education*. New York: McGraw-Hill Book Co., Inc., 1965.

Haggard, E. A. "On the Application of Analysis of Variance to GSR Data: I. The Selection of an Appropriate Measure." *Journal of Experimental Psychology*, **39** (1949), 378–92.

Haggard, E. A. "On the Application of Analysis of Variance to GSR Data: II. Some Effects of the Use of an Inappropriate Measure." *Journal of Experimental Psychology*, **39** (1949), 861–67.

Haggard, E. A. *Intraclass Correlation and the Analysis of Variance*. New York: The Dryden Press, Inc., 1958.

Hill, A. B. *Principles of Medical Statistics*. New York: Oxford University Press, Inc., 1961.

Hoel, P. G. *Elementary Statistics*. New York: John Wiley & Sons, Inc., 1966.

Kattsoff, L. O., and A. J. Simone. *Finite Mathematics with Applications in the Social and Management Sciences*. New York: McGraw-Hill Book Co., Inc., 1965.

Kerlinger, F. N. *Foundations of Behavioral Research*. New York: Holt, Rinehart and Winston, 1964.

Lacey, J. I. "The Evaluation of Autonomic Responses: Toward a General Solution." *Annals of the New York Academy of Sciences*, **67** (November 1956), 123–64.

Lathrop, R. G. "First Order Response Dependencies at a Differential Brightness Threshold." *Journal of Experimental Psychology*, **72** (1966), 120–24.

Lathrop, R. G. "Perceived Variability." *Journal of Experimental Psychology*, **73** (1967), 498–502.

Lindquist, E. F. *A First Course in Statistics*. Cambridge: The Riverside Press, 1942.

Lindquist, E. F. *Design and Analysis of Experiments in Psychology and Education*. Boston: Houghton Mifflin Co., 1956.

McCarthy, P. J. *Introduction to Statistical Reasoning*. New York: McGraw-Hill Book Co., Inc., 1957.

McGuigan, F. J. *Experimental Psychology*. Englewood Cliffs, N.J.: Prentice-Hall, Inc., 1960.

McNemar, Q. *Psychological Statistics*, 3rd Ed. New York: John Wiley & Sons, Inc., 1962.

Mood, A. M., and F. A. Graybill. *Introduction to the Theory of Statistics*. New York: McGraw-Hill Book Co., Inc., 1963.

Myers, J. L. *Fundamentals of Experimental Design*. Boston: Allyn and Bacon, 1966.

Otis, A. S. *Statistical Method in Educational Measurement*. New York: World Book Co., 1926.

Pearson, E. S., and H. O. Hartley. *Biometrika Tables for Statisticians* (Vol. I), 2nd Ed. Cambridge: Cambridge University Press, 1958.

Peatman, J. G. *Descriptive and Sampling Statistics*. New York: Harper and Brothers, 1947.

Peatman, J. G. *Introduction to Applied Statistics*. New York: Harper and Row, Publishers, 1963.

Peters, C. C., and W. R. Van Voorhis. *Statistical Procedures and Their Mathematical Bases*. New York: McGraw-Hill Book Co., Inc., 1940.

Ray, W. S. *An Introduction to Experimental Design*. New York: The Macmillan Co., 1960.

Riggs, D. S. *The Mathematical Approach to Physiological Problems*. Baltimore, Md.: The William and Wilkins Co., 1963.

Rosenthal, R. *Experimenter Effects in Behavioral Research*. New York: Appleton-Century-Crofts, 1966.

Runyon, R. P., and A. Haber. *Fundamentals of Behavioral Statistics.* Reading, Mass.: Addison-Wesley Publishing Co., 1967.

Scott, W. A., and M. Wertheimer. *Introduction to Psychological Research.* New York: John Wiley & Sons, Inc., 1964.

Sidowski, J. B. *Experimental Methods and Instrumentation in Psychology.* New York: McGraw-Hill Book Co., 1966.

Siegel, S. *Nonparametric Statistics for the Behavioral Sciences.* New York: McGraw-Hill Book Co., 1956.

Tate, M. W. *Statistics in Education and Psychology.* New York: The Macmillan Co., 1965.

Walker, H. M., and J. Lev. *Elementary Statistical Methods.* New York: Henry Holt and Co., 1958.

Wentworth, G., and D. E. Smith. *Trigonometric and Logarithmic Tables.* Boston: Ginn & Co., 1914.

Wert, J. E., C. O. Neidt, and J. S. Ahmann. *Statistical Methods in Educational and Psychological Research.* New York: Appleton-Century-Crofts, Inc., 1954.

Whitney, D. R. *Elements of Mathematical Statistics.* New York: Henry Holt and Co., 1959.

Wilks, S. S. *Elementary Statistical Analysis.* Princeton: Princeton University Press, 1956.

Winer, B. J. *Statistical Principles in Experimental Design.* New York: McGraw-Hill Book Co., 1962.

Young, R. K., and D. J. Veldman. *Introductory Statistics for the Behavioral Sciences.* New York: Holt, Rinehart and Winston, 1965.

Zimmer, H., and R. J. Krusberg. *Psychophysiologic Components of Human Behavior: A Compendium.* Athens, Ga.: University of Georgia Press, 1966.

Index

α (alpha) error, 101
ALS, 246, 256
analog measures, 238
analysis of variance, 124, 144, 230
arithmetic mean, 54
asymptote, 83
autonomic lability score, 246, 256
average, 54

bar graph, 19
β (beta) error, 103
biased estimate, 67
bimodal distribution, 79
binomial distribution, 43
biserial correlation, 191

χ^2 (chi square), 211
class interval, 53
Cochran's test, 140
coding, 71
coefficient of concordance, 204
combination, 41
conditional probability, 39
confidence limits, 102
continuous variable, 3
correction for continuity, 217
correlation, 12, 165
correlation ratio, 184
cumulative frequency histogram, 27
curvilinearity, 17

decile, 29
degrees-of-freedom, 55
dependent variable, 3
descriptive statistics, 12
deviation, 57
discrete variable, 3

errors (alpha, beta), 101, 103
η (eta), 194
expected frequency (chi square), 216

F distribution, 114
F ratio, 136
factorial design, 145

Fisher exact probability test, 222
frequency distribution, 17
frequency histogram, 19
frequency histogram (cumulative), 27
frequency polygon, 21

geometric mean, 116
goodness-of-fit, 217
graphs, 19
grouped frequency distribution, 19

histogram, 19
homogeneity of variances, 137
hypothesis testing, 96, 108

I.B.I., 254
independence (chi square), 215
independent variable, 3
inferential statistics, 12
interaction, 154, 158
interquartile range, 66
interval scale, 4

Kendall's W, 204
kurtosis, 82

λ (lambda), 249
least squares, 171, 179
leptokurtosis, 82
level of significance, 99
linearity of regression, 171
logarithms, 245

Mann-Whitney U Test, 224
mathematical proof, 5
mean, 54
mean squared deviation, 58
median, 52
mesokurtosis, 82
mode, 49
μ (mu), 54
multimodal distribution, 79

nominal scale, 4
normal distribution, 82

303

null-hypothesis, 99, 120

one-tail test, 102
ordinal scale, 4
ordinate, 20

parameter, 8
parametric assumptions, 98
parametric measures, 10
percentiles, 27
permutations, 39
phi (ϕ) coefficient, 200
platykurtosis, 82
point biserial correlation, 193
point estimate, 105
populations, 8
probability, 34

quantiles, 27
quartiles, 29

random sample, 8
range, 15
rank-order correlation, 201
ratio scale, 4
regression, 177
rejection region, 103

samples, 8
sampling distribution of $\overline{X}$, 86
scaling, 3
scatter diagram, 168
sequential dependency, 249
sign test, 223
significance of differences, 100
skewness, 79
Spearman's rho (ρ), 202

t, 90
tetrachoric correlation, 198
treatment group, 125
trend analysis, 230
two-tail test, 102

U test, 224

variability, 63
variance, 66

X axis (abscissa), 20

Y, 82
Y axis (ordinate), 20

z, 74
Z, 84